Progress in Molecular and Subcellular Biology 9

9 Progress in Molecular and Subcellular Biology

With Contributions by
M. Bachmann, C. Brunel, P. Jeanteur
R. Messer, W. E. G. Müller, R. E. Rhoads
H. C. Schröder, J. Sri-Widada

Edited by F. E. Hahn, D. J. Kopecko
W. E. G. Müller

Managing Editor: F. E. Hahn

Springer-Verlag
Berlin Heidelberg New York Tokyo

Professor Fred E. Hahn, Ph. D.
Division of Communicable Diseases and Immunology
Walter Reed Army Institute of Research
Washington, D. C. 20307, USA

With 37 Figures

ISBN-13: 978-3-642-70205-1 e-ISBN-13: 978-3-642-70203-7
DOI: 10.1007/978-3-642-70203-7

Library of Congress Catalog Card Number 75-79748

Softcover reprint of the hardcover 1st edition 1985

2131/3130-543210

Contents

W.E.G. MÜLLER: Cell-Cell Recognition in Lower Multicellular Eukaryotes. Models Supporting the Modulation Theory of Adhesion (With 9 Figures)

Contributors

BACHMANN, M., Institut für Physiologische Chemie, Abt. "Angewandte Molekularbiologie", Universität, Duesbergweg, 6500 Mainz, FRG

BRUNEL, C., Laboratoire de Biochimie, Centre Paul Lamarque BP 5054, 34033 Montpellier and Laboratoire de Biologie Moléculaire, Université des Sciences et Techniques du Languedoc, 34060 Montpellier, France

JEANTEUR, P., Laboratoire de Biologie Moléculaire, Université des Sciences et Techniques du Languedoc, 34060 Montpellier, France

MESSER, R., Institut für Physiologische Chemie, Abt. "Angewandte Molekularbiologie", Universität, Duesbergweg, 6500 Mainz, FRG

MÜLLER, W.E.G., Institut für Physiologische Chemie, Abt. "Angewandte Molekularbiologie", Universität, Duesbergweg, 6500 Mainz, FRG

RHOADS, R.E., Department of Biochemistry, University of Kentucky Medical Center, Lexington, KY 40536-0084, USA

SCHRÖDER, H.C., Institut für Physiologische Chemie, Abt. "Angewandte Molekularbiologie", Universität, Duesbergweg, 6500 Mainz, FRG

SRI-WIDADA, J., Laboratoire de Biochimie, Centre Paul Lamarque BP 5054, 34033 Monpellier and Laboratoire de Biologie Moléculaire, Université des Sciences et Techniques du Languedoc, 34060 Montpellier, France

snRNP's and scRNP's in Eukaryotic Cells

C. Brunel, J. Sri-Widada and P. Jeanteur

A. Historical Background

The first low mol. wt. RNA species different from ribosomal 5S and 5.8S RNA's and tRNA were the U snRNA's described more than 15 years ago in eukaryotic cells nucleoli (Muramatsu et al. 1966) and nuclei (Larsen et al. 1967; Weinberg and Penman 1968) and were recognized as discrete, abundant, and stable species. Their cellular distribution, synthesis, sequence analysis, and association with other nuclear macromolecules or substructures were investigated. Yet their function remains mostly a matter of speculation. However, an outsurge of interest was aroused by two simultaneous breakthrough observations made within the last years.

The first one was the recognition of a striking base pairing capability between the 5' end of U_1 snRNA, which has been highly conserved during evolution, and the intron side of splice junctions which also have a few consensus nucleotides at both their 5' (acceptor) and 3' (donor) end. A similar complementarity between adenovirus-2 VA RNA and a splice junction of the late adenovirus-2 transcript has also been noticed (Murray and Holliday 1979). On this basis, it has been proposed that U_1 snRNA could play a role in positioning intron-exon junctions (Rogers and Wall 1980; M.R. Lerner et al. 1980) in proper register for accurate splicing. Proteins specifically associated with U_1 are likely to participate in this recognition as well as in the splicing process itself.

The second one was the discovery that antibodies from patients suffering from autoimmune diseases, like Systemic Lupus Erythematosus (SLE), recognize the U snRNA's (except U_3) in association with a set of proteins (M.R. Lerner and Steitz 1979) which were immediately characterized as being neither histones nor proteins known to be associated with hnRNA in hnRNP. The impact of this finding has been enormous and opened new avenues of research along at least three major lines: firstly, it provided invaluable tools for probing both the structure and function of snRNP's; secondly, it provided important cues for the understanding of autoimmune diseases as well as a basis for a molecular approach of their pathology and pathogenesis; last but not least, it prompted an intensive investigation of other autoimmune antibodies which soon led to the discovery of new families of antigenic RNP's, many of which were found in the cytoplasm.

In this review, we will restrict ourselves to the low molecular RNA's which are found as RNP's. This will exclude the small nu-

2

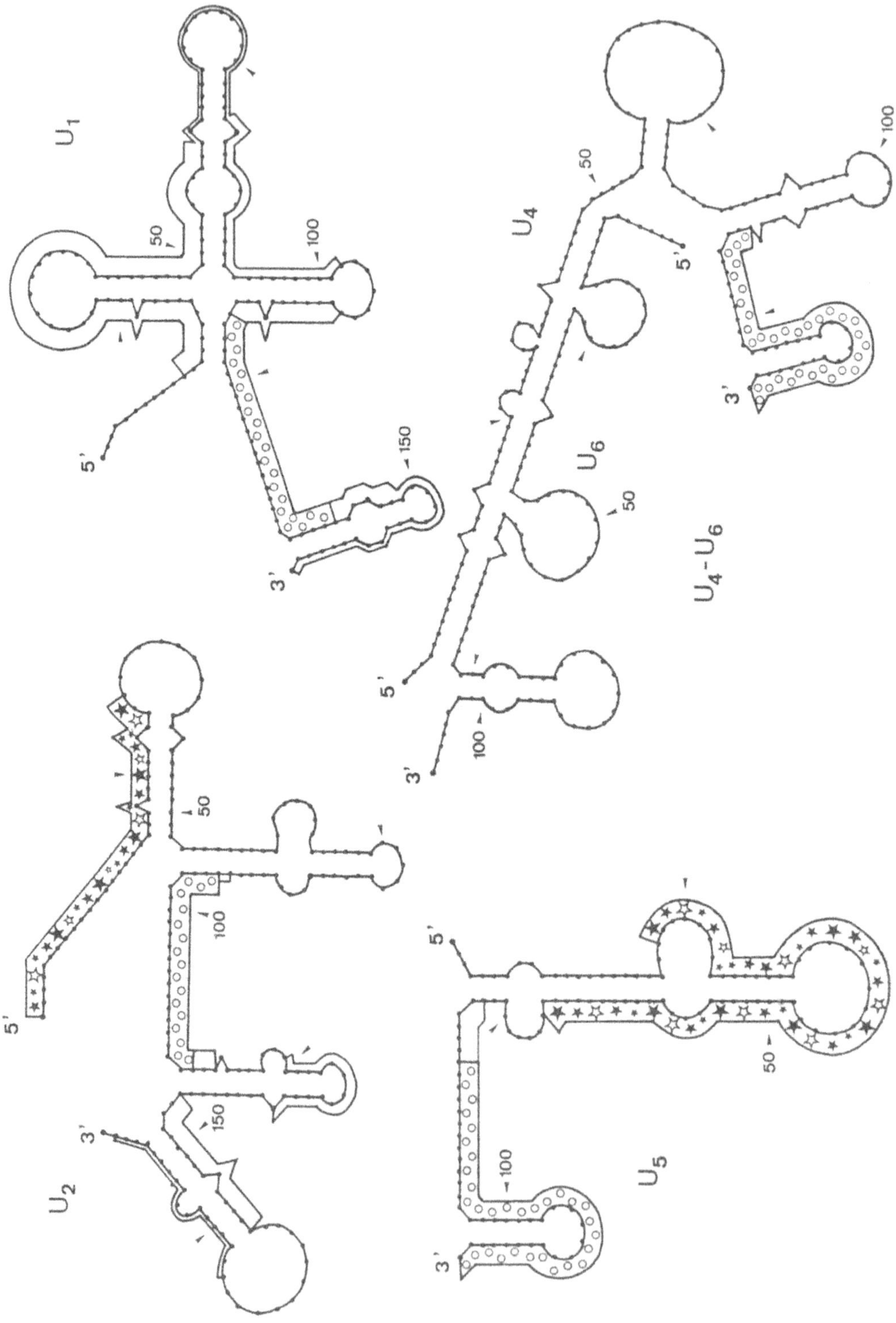

clear RNA's transcribed by RNA-polymerase I (Benecke and Penman 1977, 1979; Reichel et al. 1982) and referred to as snP1 RNA's. We will also overlook the entire field of translational effectors RNA's (Kennedy et al. 1978; Pluskal and Sarkar 1981).

B. The U snRNA's

In this chapter, our goal is to review recent results in the field of isolation, structure, and function of U snRNP's. Detailed accounts of the structural aspects of U snRNA's have already been published in recent reviews (Hellung-Larsen 1977; Reddy and Busch 1981; Zieve 1981; Busch et al. 1982), so that they will only be briefly summarized here.

I. Primary and Secondary Structures of U snRNA's

First recognized on the basis of their sedimentation and electrophoretic properties, the small nuclear RNA's were named U snRNA's on account of their relatively high content in uridylic acid residues. Several reviews (see above) have already been dedicated to this rather historical aspect which will intentionally not be covered here.

Actually six major species of these snRNA's are known and numbered U_1 to U_6. They are abundant (10^6 molecules by cell for U_1 and from 10^4 to 5×10^5 molecules by cell for the others) (Weinberg and Penman 1969; Marzluff et al. 1975), metabolically stable (half-lives of up to one cell cycle) (Busch et al. 1982) and were identified in all animal cells from insect to human. They are also present in dinoflagellates (Reddy et al. 1983a) and in plants (Krol et al. 1983). The 5' ends of U_1 to U_5 snRNA's are all capped by trimethylguanosine (Busch et al. 1982) while the nature of the modification of the of U_6 has not yet been completely identified

---

◄ Fig. 1. Localization of protected snRNA sequences within core snRNP's, hnRNP's and "native" snRNP's.
Lines with circles refer to core snRNP sequences protected against micrococcal nuclease (Liautard et al. 1982). *Star lines* refer to sequences protected in hnRNP's (Sri-Widada et al. 1983). *Blank lines* refer to sequences protected in "native" snRNP's (Reveillaud et al. 1984; Lelay-Taha et al. submitted). *Black points* represent nucleotides: The idealized schemes are based on, the following published primary and secondary structure data.
U_1: Branlant et al. 1980, 1981. Mount and Steitz 1981.
U_2: Branlant et al. 1982 and pers. Comm.; Reddy et al. 1981a.
U_4 and U_5: Krol et al. 1981a,b.
U_6: Harada, Kato and Nishimura 1980.
The model showing the $U_4 - U_6$ interaction is from Hashimoto and Steitz (1984). For other published primary and secondary structures, see also: Epstein et al. 1980; Kato and Harada 1981a,b; Mount and Steitz 1982; Reddy, Henning and Busch 1980, 1981; Reddy et al. 1974a-c; Shibata et al. 1975; Wise and Weiner 1980; Myslinski et al. 1984

(Epstein et al. 1980). All contain modified nucleosides. The primary sequence of all U snRNA's are known in several species and secondary structures have been worked out as well for some of them (Fig. 1).

II. Subcellular Localization and Synthesis

The subcellular localization of U snRNA's (especially U_3, U_2, and U_1) has been a matter of controversy. snRNA's were originally found in the nucleus and claimed to be absent from the cytoplasm (Muramatsu et al. 1966; Larsen et al. 1967, 1969; Weinberg and Penman 1968; Dingman and Peacock 1968). Several investigators have then reported the presence of significant amounts of U_2 and U_1 in the cytoplasm (Frederiksen and Hellung-Larsen 1975; Zieve and Penman 1976). Conflicting results have been also obtained with nonaqueous methods of cell fractionation. When lyphilized cells were homogenized and centrifuged in anhydrous glycerol (Eliceiri and Gurney 1978; Gurney and Eliceiri 1980). snRNA species U_3, U_2, and U_1 were almost entirely (> 95%) nuclear. In contrast, Frederiksen et al. (1981) have found considerable amounts of U_2 (57%) and U_1 (43%) snRNA's in cytoplasmic fractions isolated by centrifugation in mixtures of cyclohexane and tetrachloromethane. It seems clear, however, that the presence of snRNA's in cytoplasmic fractions is either transient or merely artifactual and it is now widely accepted that their function, whatever it is, takes place in the nucleus where their mature, steady-state form belongs in vivo. As to U_3 snRNA, its stable localization is clearly nucleolar (Busch et al. 1982).

Notwithstanding their final cellular site of accumulation, at least some snRNA's or their precursors may transiently appear in the cytoplasm. U_3 is found in the cytoplasm within the first few minutes after its transcription, then equally distributed in the cytoplasm and the nucleolus 2 h later and finally almost entirely in the nucleolus after 16 h (Zieve and Penman 1976). Long precursor molecules of U_2, U_1, and U_4 have been detected in the cytoplasm (Eliceiri 1974; Frederiksen and Hellung-Larsen 1975; Zieve and Penman 1976; Eliceiri and Sayavedra 1976; Eliceiri and Gurney 1978; Eliceiri 1979, 1980). As they are recognized by both autoimmune sera and monoclonal Sm antibody (Chandrasekharappa et al. 1983; Madore et al. 1984a,c) these precursor molecules, between one and at least eight nucleotides longer than mature RNA in the case of U_1 (Madore et al. 1984a), assemble with proteins and then are rapidly processed at their 3' end prior to migrating into the nucleus (see Sect.B. VI.8. below). RNA's whose properties are compatible with being U_3 RNA precursors have been also detected (Chandrasekharappa et al. 1983).

III. In Vivo Transcription of U snRNA's

Controversial results have been reported about the type of RNA polymerase involved in U snRNA's synthesis. Using inhibitors like α-amanitin, actinomycin D1, camptothecin, and toyocamycin, Zieve et al. (1977) suggested that RNA polymerase I might be responsible for the synthesis of U_3, U_2, U_1, U_4, and U_6. As of now, there has been no confirmation of this claim and the debate is more likely restricted to RNA polymerases II and III.

Strong but not quite final evidence suggests that RNA polymerase II, which is clearly established as responsible for the synthesis of hnRNA, also synthesizes U snRNA's (at least U_2, U_3, and U_1) in agreement with the first observation by Ro-Choi et al. (1976). These include: (a) The sensitivity of U_3, U_2, and U_1 snRNA's synthesis to low concentrations of α-amanitin in whole cell systems (Frederiksen et al. 1978; Ro-Choi et al. 1976, Gram Jensen et al. 1979; Eliceiri 1980; Tamm et al. 1980) as well as, in the case of U_1, in isolated nuclei (Roop et al. 1981) or in cell-free systems (Murphy et al. 1982); (b) The inhibition of the synthesis of U_3, U_2, and U_1 at high temperature in a mutant cell line of baby hamster kidney carrying a temperature-sensitive polymerase II (Hellung-Larsen et al. 1980); (c) The relative sensitivity of U_3, U_2, and U_1 snRNA's synthesis to DRB (5,6 dichloro-1-βD-ribofuranosyl benzimidazole) (Hellung-Larsen et al. 1981), a nucleoside analogue known to promote premature termination of hnRNA transcription (Tamm and Kikuchi 1979). No information pertaining to the synthesis of U_4, U_5, and U_6 species in particular is available yet.

Despite the above evidence for the involvement of RNA polymerase II there are at least three kinds of contradictory arguments: (1) Earlier experiments on isolated nuclei (Udvardy and Seifart 1976) and on nuclear preparations supplemented with exogenous RNA polymerase III (Sklar and Roeder 1977) suggested that this latter enzyme catalyzes the synthesis of U_3, U_2, and U_1 snRNA's and this was also concluded from experiments with mouse myeloma cells (Brown and Marzluff 1978); (2) The presence within U_1 of short sequences highly reminiscent of RNA polymerase III transcription initiation signals (Galli et al. 1981); (3) The existence of precursors which, if confirmed, would lead to the unprecedented situation of transcription initiation sites being clearly different from cap sites. In this respect, as well as in others which will be discussed later, it should be pointed out that U_6 seems to stand apart from the other U snRNA's.

IV. Genes and Pseudogenes for U snRNA's

Clones of genomic DNA from various species containing sequences at least partially complementary to U_1 (Roop et al. 1981; Denison et al. 1981; Van Arsdell et al. 1981; Westin et al. 1981; Murphy et al. 1982; Denison and Weiner 1982; Monstein et al. 1982; Manser and Gesteland 1982; Piechaczyk et al. 1982); Lund and Dahlberg 1984; Zeller et al. 1984), U_2 (Denison et al. 1981; Van Arsdell et al. 1981; Westin et al. 1981; Alonso et al. 1983; Van Arsdell and Weiner 1984a), U_3 (Wise and Weiner 1980; Denison et al. 1981; Van Arsdell et al. 1981; Berstein et al. 1983; Marzluff et al. 1983), U_4 (Hammarstrom et al. 1982) and U_6 (Ohshima et al. 1981b; Hayashi 1981) are now available. All earlier reports agree with the conclusion that these sequences, whether true genes or pseudogenes, are abundant, not clustered or in small tandem repeats but, rather, scattered thoughout the genome. However, it appears from several recent reports that genes, especially true genes, are tandemly repeated. In sea urchin, genes coding for U_1 and U_2 snRNA's are in distinct tandemly repeated clusters of at least 30 kb with a repeat length of 1100 - 1400 base pairs (Card et al. 1982). In *Xenopus laevis*, the U_1 and U_2 genes are organized in tandemly repeated units of 830 bp and 1.8 kb respectively for U_2 and U_1 (Mattaj and Zeller 1983; Zeller et al. 1984). In *D. Melanogaster*, the number of U_2 snRNA genes is very small (four to five) and this seems also to be true for U_1 snRNA (Alonso et al. 1983). In man, genes for U_2 snRNA are organized as a nearly perfect tandem array of 10 to 20 copies per haploid genome, with a basic repeating unit of 6 kb (Van Arsdell and Weiner 1984a; Westin et al. 1984).

Attempts at localizing U snRNA's genes at the chromosomal level are just being started. Using cell fusion techniques, all true human U_1 genes, i.e., producing authentic U_1 snRNA, were found to be localized on chromosome 1 (Lund et al. 1983). By in situ hybridization to the chromosomes of *Drosophila* salivary glands, Saluz et al. (1983) were able to accurately localize genes for U_2, U_3, U_4, and U_6 on specific bands. The number of hybridizing sites (about three) correlates well with the number of genes estimated by Southern-blot analysis.

Clones containing sequences perfectly matching the sequence of the homologous U_1 snRNA were obtained from several species: chicken (Roop et al. 1981), rat (Watanabe-Nagasu et al. 1983), and man (Murphy et al. 1982; Monstein et al. 1982; Manser and Gesteland 1982; Lund and Dahlberg 1984). Sequences more or less closely related to the canonical TATA box were found upstream from the 5' end of the coding sequence in only two (Monstein et al. 1982; Lund and Dahlberg 1984) of the human U_1 clones studied so far, but also in clones for U_6 (Ohshima et al. 1981b), D_2 which is the *Dictyostelium* equivalent of U_3 (Wise and Weiner 1980) and U_1, U_2 of *D. Melanogaster* (Beck et al. 1984). In the case of one human U_1 clone (Murphy et al. 1982), transcription of a 592 bp fragment was analyzed both in vivo and in vitro. Deletions of 5' flanking sequences have demonstrated that the DNA sequence required for in vivo transcription is located at least 100 nucleotides upstream from the beginning of the coding sequence while in vitro transcription in a HeLa cell extract containing RNA polymerase II (Manley et al. 1980) yielded a molecule starting 183 nucleotides upstream of the capping site, thereby suggesting that U_1 snRNA might derive from a larger precursor. Coherent with this finding is recent work from Van Arsdell and Weiner (1984a), who identified two regions of strong homology in human U_1 and U_2 genes. Region II seems to correspond to that found by Murphy et al. 1982. However, it is located 174 bp further upstream from the 5' end of the snRNA sequence in U_1 gene that in U_2 gene, suggesting that its position relative to the initiation site can vary.

Numerous clones have been characterized as pseudogenes which, as pointed out by Denison and Weiner (1982) can include scattered base substitutions, short internal deletions and loss of sequences corresponding to the 3', but never the 5', end of the mature snRNA's. They are scattered throughout the genome in an approximate ratio of 10 to 1 as compared to the true genes.

According to Denison and Weiner (1982), three classes of U_1 pseudogenes have been distinguished. Class I pseudogenes have considerable flanking sequence homology with the authentic U_1 gene and were probably derived from it by gene duplication. Class II and III pseudogenes have flanking sequences unrelated to those of the true gene and short direct repeats flank class III pseudogenes. It was then proposed that class II and III pseudogenes were generated by a RNA-mediated mechanism involving the insertion of U_1 sequence information into a new chromosome locus either as a cDNA copy or as the RNA itself. Similar propositions were made recently for human U_2 (Van Arsdell and Weiner 1984b) and U_3 (Bernstein et al. 1983) pseudogenes which could be generated in vivo by integration of a self-primed cDNA copy of U_3 snRNA at new chromosomal sites.

V. The RNP Status of U snRNA's

All U snRNA's appear not to be free in the nucleus but rather directly associated with specific proteins.

1. Earlier Reports on the Existence of snRNP's

The first indication of a possible RNP structure came from an earlier report by Enger and Walters (1970) demonstrating that small methylated RNA molecules found in the nucleus sedimented more rapidly (10-30S) than the corresponding naked RNA's (5-10S). Rein (1971), studying the snRNA's in cells blocked in mitosis by colcemide, showed that they sedimented as 30 to 180S structures which were partially dissociated by 0.3 M NaCl. In neither case, however, was the nature of these structures further characterized. Raj et al. (1975) were the first to demonstrate that snRNP's represent a real association between snRNA's (in this case U_1 and U_2) and proteins in the form of an RNP structure sedimenting at about 20S. Similar results were obtained by Howard (1978), who isolated by sedimentation of a nucleoplasmic extract of mouse erythroleukemia cells a 10S particle containing snRNA's whose density (1.38 g cm^{-3}) in cesium chloride suggested an RNP nature.

2. Antigenicity of snRNP's versus Natural Autoimmune Antibodies

It has been known for a long time that autoimmune rheumatic diseases like Systemic Lupus Erythematosus (SLE) are characterized by antinuclear antibodies (ANA). Antibodies with specificities against DNA, histones, and other structures have been recognized in sera from these patients. The nature of these antibodies and their immunological as well as clinical significance have been extensively dealt with in two exhaustive reviews by Tan (1979, 1982) to which the reader is referred for a more detailed account. Among this wide class of autoimmune rheumatic disorders, a new distinct clinical entity called mixed connective tissue disease (MCTD) has been described by G.C. Sharp et al. (1972). High titers of an antibody against a nuclear antigen sensitive to both RNAse and trypsin (Northway and Tan 1972) and therefore

Table 1. Proteins associated with U snRNA's

Method		Cell	snRNA	Proteins		Reference
				M.W. < 20 k	M.W. > 20 k	
Immuno-precipitation with	Anti-Sm	Ehrlich ascites	$U_1 - U_6$	11 k, 12 k, 13 k, 16 k	22 k, 28 k, 33 k	(a)
	Anti-RNP		U_1			
	Anti-SM	F. eryth. leuk.	$U_1 - U_2$	3(13 k), 16 k	20 k, 25 k, 26 k, 31 k	(b)
	Anti-RNP		U_1			
	Anti-Sm	HeLa	$U_1 - U_6$	10 k, (12–14 k)	26 k, 30 k	(c)
	Anti-RNP		U_1			
	Anti-Sm	HeLa	$U_1 - U_6$	4(9–14 k)	20 k, 25 k, 28 k	(d)
	Anti-RNP		U_1			
	Anti-(U_2)RNP	HeLa	U_2	11 k, 12 k, 13 k, 16 k	28 k, 32 k	(o)
Immuno-affinity columns on		Rat liver	N.D.	13 k	30 k	(e)
	Anti-RNP	Calf thymus	N.D.			
	Anti-Sm			5(12–13 k)	2(30 k), 65 k	(f)
		Calf thymus	N.D.			
	Mixed	Rabbit thymus	(U_2-U_1) ?	10 k, 11 k, 13 k	25.5 k, 27.5 k	(g/h)
	Anti-Sm/ Anti-RNP				31 k, 35 k, 73 k	

Solid phase radioimmunoassay revealed with	Anti-Sm	Calf thymus		13 k		(f)
	Anti-RNP				30 k, 65 k	
Protein blots revealed with	Anti-RNP	Rabbit thymus			40 k, 2(70 k)	(i)
		Rat liver			30 k	(j)
	Anti-Sm	Rat liver		13 k		(j)
		Rabbit thymus		13 k	2(16–27 k)	(h)
		HeLa				
	Anti-RNP and Anti-RNP (monoclonal)	Rabbit thymus			73 k	(h)
		HeLa			69 k	
	Anti-(U_2)RNP	HeLa			32 k	(o)
Biochemical purification of snRNPs	CsCl+sarcosyl	HeLa	U_1	4/5(9–14 k)		(k)
	Centrifugation		U_1 to U_6	4/5(9–14 k)		(l)
	Ion exchange chromatography	F.eryth.leuk.	U_1	11 k, 12 k, 13 k, 16 k	22 k, 28 k, 33 k, 68 k	(m)
		HeLa			22 k, 28 k, 29 k, 68 k	
		id	$U_2U_4U_5U_6$	11 k, 12 k, 13 k, 16 k	28 k, 32 k	(m)
		HeLa	U_1	8.5 k, 9.1 k, 10.2 k 12.3 k, 17.5 k	21.5 k, 23 k, 30 k 67 k	(n)
		id	U_2	8.5 k, 9.1 k, 10.2 k, 12.3 k	21.5 k, 23 k, 27 k	(n)

(a) Lerner and Steitz, 1979; (b) Matter et al., 1982; (c) Barque et al., 1981; (d) Assens et al., 1982a; (e) Douvas et al., 1979; (f) Takano et al., 1981; (g) White, Gardner and Hoch, 1981; (h) Billings and Hoch, 1983; (i) White and Hoch, 1981; (j) Douvas, 1982; (k) Sri-widada et al., 1982; (l) Brunel et al., 1981; (m) Hinterberger, Peterson and Steitz, 1983; (n) Kinlaw, Robberson and Berget, 1983; (o) Mimori et al., 1984.

designated RNP, are usually found in this disease, although they are not strictly diagnostic for it. This RNP antigen has been isolated from rat liver nuclei and characterized using affinity columns containing the antibody (Douvas et al. 1979). Two major polypeptides of mol. wt. 30 k and 13 k were identified in the protein moiety of this RNP. However, the RNA complement was not identified as snRNA's at this time. Another nuclear antigen, called Sm and characterized by its relative resistance to both trypsin and RNAse, had been identified even earlier and shown to be diagnostic for SLE (Tan and Kunkel 1966).

An important breakthrough came with the discovery by M.R. Lerner and Steitz (1979) that both anti-RNP and anti-Sm antibodies from SLE patients precipitated snRNP's from Ehrlich ascites tumors and HeLa cells. They definitively established that the anti-RNP antibody is directed against U_1 snRNP and accordingly proposed (M.R. Lerner and Steitz 1981) to rename it anti-U_1 RNP. Seven polypeptides with mol. wts. ranging from 12 to 35 k, designated A to G were found to be immunoprecipitated along with U_1 snRNA. Anti-Sm immunoprecipitated all the U_2, U_1, U_4, U_5, and U_6 snRNA's and the same seven polypeptides were found in the precipitate indicating that all these U snRNA's have a common set of associated proteins. This was later confirmed using a monoclonal antibody (E.A. Lerner et al. 1981). These proteins are clearly distinct from either histones or the major hnRNP polypeptides (M.R. Lerner and Steitz 1979). Two new sera have been described recently. The first one, discovered in a patient with scleroderma-polymyositis overlap syndrome, contains antibodies directed against U_2 snRNP. Along with the U_2 snRNA, six polypeptides are precipitated among which is a U_2 specific component (Mimori et al. 1984) (see Sect. V.5 and Table 1). The second one was found on scleroderma patients and recognizes, on protein blots, a 34 k component probably associated with U_3 snRNA (Reddy et al., pers. comm.). This seems to be the first demonstration that U_3 snRNA exist in a snRNP form.

3. Evidence for A core snRNP Structure

At the same time when immunological methods were introduced, a strictly biochemical approach to the purification of snRNP's was being developed in our laboratory. This procedure starts with purified hnRNP's to which at least a major fraction of snRNA's and, therefore, snRNP's are bound (see below). A ribonucleoprotein complex whose RNA complement consists exclusively of small nuclear RNA's has been purified from particles containing heterogenous nuclear RNA (hnRNP) from HeLa cells. This was accomplished by taking advantage of their ability to band at a density of about 1.43 g cm^{-3} in plain cesium chloride, as well as in cesium chloride gradients containing 0.5% sarcosyl without prior aldehyde fixation. After these two steps of equilibrium density centrifugation, these snRNP's were still largely contaminated by free proteins (and especially phosphoproteins). A final step of purification by velocity sedimentation in a sucrose gradient containing 0.5% M salt and 0.5% sarcosyl was efficient in completely eliminating all free proteins. The sedimentation coefficient measured in the presence of sarcosyl has been shown to be 11 to

12 S. All U snRNA's. U_2, U_1, U_4, U_5, and U_6, were found in these purified snRNP's, although the recovery was not as good for all of them, that of U_6 being noticeably lower. Electrophoresis of associated proteins revealed four of five bands with migration corresponding to mol. wt. ranging between 9 and 14 k (Brunel et al. 1981; Sri-Widada et al. 1981).

A further step was taken by the purification of the individual U_1 snRNP (Sri-Widada et al. 1982). This was obtained by combining the above procedure with the preferential release of this particle from hnRNP's under isotonic conditions as observed by other authors (Gallinaro and Jacob 1979; Zieve and Penman 1981). This U_1 snRNP contained no other snRNA and had a protein complement identical to that of the mixture of all snRNP's (Brunel et al. 1981). This finding did away with the possibility that each snRNA could be bound to only one of these small mol. wt. polypeptides and demonstrated that all individual snRNP's isolated by this procedure have the same protein composition and therefore that the specificity of their function depends only on their snRNA component and/or their cognate proteins which have been lost during the purification. Our rather drastic purification procedure has therefore yielded a "core particle" structure which is common to all nucleoplasmic snRNP's. Due to their simplified protein complement, these core particles can be expected to be more easily amenable to structural studies.

Although certainly far from being native, these core snRNP's retain an important part (46%) of their antigenicity towards anti-Sm antibodies (Assens et al. 1982b). Moreover, digestion of a great part of the RNA moiety does not abolish this antigenicity. This indicates that at least some of the Sm antigenic determinants are among the 4 - 5 small polypeptides with mol. wt. between 9 and 14 k, as confirmed by electroblotting techniques in which a 13 k protein is shown to react with anti-Sm (White and Hoch 1981; Takano et al. 1981). The antigenicity towards RNP antibody is nearly entirely lost in "core snRNP's" indicating that the RNP determinant must depend on proteins specifically interacting with U_1 and which are lost during the purification. The previously suggested (Takano et al. 1981) presence of RNP determinants among the high mol. wt. proteins was also confirmed by others (White and Hoch 1981; Billings et al. 1982) using electroblotting techniques (see below).

4. Attempts at Isolating Native snRNP's

Several recent reports were aimed at purifying snRNP's away from other cellular components on one hand and at separating individual particles on another hand. These procedures were based either on conventional fractionation methods or on the use of antibodies against the unusual cap ($m_3{}^{2,2,7}G$) which is found only in U snRNA's except U_6.

Along the first line, a procedure was developed by Hinterberger et al. (1983) to fractionate snRNP's from mouse Friend erythroleukemia and human HeLa cells. A nuclear extract obtained at 0.4 M salt was submitted to a series of gel filtration, ion exchange and hydrophobic/ion exchange chromatographic steps. Three final fractions were obtained: the first one contained only U_1 snRNP, the second one was a mixture of $U_1 + U_2$ snRNP's and third one contained essentially U_2 snRNP with minor amounts of all other snRNP's. These partially purified snRNP's were stable and retained their immunoprecipitability by anti-Sm or anti-RNP antibodies. Another report came from Kinlaw et al. (1983) which used a protocol quite similar to the above one. By submitting a mixture containing all but U_5 snRNP's successively to heparin-agarose, ω-aminopentyl agarose, and cibacron blue chromatographic steps, they succeeded in obtaining a fraction containing exclusively U_2 snRNP. A protocol to simultaneously isolate U_1 and U_5 snRNP's has been developed in our laboratory (Lelay-Taha et al., submitted). It is based on our demonstration (Reveillaud et al. 1984) that Mg^{2+} stabilizes RNA-protein interactions in snRNP's, allowing their purification in cesium chloride.

Along the second line, recent reports (Bringman et al. 1983a,b; Smith and Eliceiri 1983) have made use of antibodies specific for 2,2,7-trimethyl guanosine ($m_3{}^{2,2,7}G$) to study the accessibility of the 5'-terminal cap of snRNA's, either as naked RNA, or in snRNP's. These antibodies migth turn out to be a useful tool for recovering snRNP's from nuclear extract by affinity chromatography. Along the same line, the procedure proposed by Wilk et al. (1982) which uses m-aminophenyl boronate subtituted agarose to specifically bind RNA chains carrying a 5' cap might also be applied to the purification of snRNP's.

5. The Protein Complement of snRNP's

Until a specific function can be assigned to each individual snRNP; thereby providing a functional assay for their integrity, delineating the snRNP entity will remain a matter of operational definition. Despites this limitation, a rather good overall consistency emerges from the comparison of the protein complements of particles prepared by quite different methods, if one allows for slight discrepancies in mol. wt. determinations made in different laboratories (Table 1).

Small polypeptides in the 9 - 14 k range (D to G in the nomenclature of M.R. Lerner and Steitz 1979) have been found in particles obtained by preparation procedures (see above) as widely different as those involving immunoprecipitation with anti-Sm and anti-RNP antibodies (M.R. Lerner and Steitz 1979) or various purification protocols (Hinterberger et al. 1983; Kinlaw et al. 1983; Bringmann et al. 1983b; Billings and Hoch 1984) including the most drastic one developed in our laboratory (Brunel et al. 1981). This constant occurence strongly suggests that core proteins exist which are common to all snRNP's. As will be discussed later, this argument is further strengthened by the observation that these proteins are extremely tightly bound to a structural domain common to all snRNA's, with the exception of U_6 (Liautard et al. 1982). These proteins appear to be basic (Brunel et al., unpublished results) on the grounds of their behavior in non-equilibrium pH gradient electrophoresis (O'Farrell et al. 1977).

At variance with the core proteins and the B B' doublet which are common to all snRNP's, at least some of the higher mol. wt. proteins are specific for a single snRNA species (Table 1). Among them, there appears that three polypeptides of 22 (C), 33 (A) and 68 k are unique to U_1 snRNP (Hinterberger et al. 1983). These same authors show that a 32 k (A') component is present only in a chromatographic fraction enriched in U_2 snRNP and recently have demonstrated by protein blot analysis that this component carries U_2 (RNP) determinant(s) (Mimori et al. 1984). Kinlaw et al. (1983) have found that U_2 snRNP contains a specific polypeptide of 27 k which is barely detectably in methionine-labeled but readily apparent in leucine-labeled snRNP's. It was called B" by Pettersson et al. (1984) and also seems to be antigenic towards anti-U_2(RNP). One polypeptide is unique to U_5 snRNP (Lelay-Taha et al., submitted). Other polypeptides seemed to be unique to U_4 snRNP and three components (23, 12 and 10 k) appeared to be resolved into multiple forms by two-dimensional electrophoresis which could be differentially represented in individual snRNP's (Kinlaw et al., pers. comm.). The question as to whether the 68 k component, which seems to be phosphorylated (Wooley et al. 1983), could be identical to a protein of the nuclear matrix (Berezney and Coffey 1974) remains open (Habets et al. 1983a).

6. Identification of Sm and RNP Antigens

The proteins found in purified snRNP's correspond well to those found by immunoprecipitation (M.R. Lerner and Steitz 1979; Barque et al. 1981; Matter et al. 1982; Assens et al. 1982b; Wieben and Pederson 1982) and immunoaffinity columns (Douvas et al. 1979; Takano et al. 1980; White and Hoch 1981; White et al. 1981) using anti-RNP and anti-Sm antibodies (see Table 1). A more direct identification of the proteins actually carrying the Sm and RNP determinants is currently emerging from the use of proteins blotting techniques. The Sm determinant(s) have been found in a polypeptide (or closely spaced doublet) of 25 to 28 k (B, B') and/or in a 13 to 16 k species (D) (Billing and Hoch 1983) depending upon the antigen source and its purification (White et al. 1981; Lerner et al. 1981; White et al. 1982; Douvas 1982). The RNP determinant was identified by White and Hoch (1981), Takano et al. (1981) and by Billings et al. (1982) using an anti-RNP monoclonal antibody. It seems to reside in a 65 k to 70 k polypeptide and perhaps in a smaller species (30 to 40 k) possibly derived by proteolysis of the former (Billings and Hoch 1983). A more complete description of the snRNP polypeptides recognized on immunoblots by patient sera is now available (Pettersson et al. 1984). It appears that the B, B', D, and in some cases E polypeptides are recognized by anti-Sm, while the 68 k, A, and/or C polypeptides are recognized by anti-RNP. To circumvent the problems of denaturation inherent in the immunoblotting techniques, Fisher et al. (1983, 1984) have analyzed neo-synthesized snRNP proteins, either generated in vitro by cell-free translation or in vivo by pulse labeling. Considering the presence, under these conditions, of unassembled snRNP proteins, they found that all the A to G polypeptides are antigenic towards anti-Sm, while A and C, which are unique to U_1 snRNP, are recognized by anti-RNP.

In order to know whether snRNP proteins display the same degree of evolutionary conservation as the snRNA's themselves, Wieben and Pederson (1982) have examined the protein composition of snRNP's from *Drosophila* using human anti-RNP antibody. Two proteins only were immunoprecipitated along with an RNA species identified as U_1 snRNA by hybridization with a clone of U_1 DNA. These comigrated in SDS gels with the human B and D polypeptides and had mol. wts. of 26 and 14 k respectively. Four additional proteins were detected with the anti-Sm antibody only two of which comigrated with human E and F proteins (about 10 k).

Using also a *Drosophila* extract, Wooley et al. (1982) identified two polypeptides of 26 and 18 k reactive towards human anti-Sm antibody by western blotting. Interestingly it seems that the 18 k protein, which does not correspond to the 14 k previously found in *Drosophila* extracts by Wieben and Pederson (1982), reacts predominantly with fractions enriched in K5 and K6 RNA species (corresponding to U_1 and U_2 snRNA's respectively), suggesting that this polypeptides is not a degradation product of the 26 k and is unique to K5 and K6.

As to the 26 k polypeptide, both reports from Wieben and Pederson (1982) and Wooley et al. (1982) agree with the suggestion that this polypeptide is an evolutionary conserved antigenic component of *Drosophila* and mammalian snRNP's.

7. Assembly of snRNP's

Two kinds of very exciting assembly experiments have recently been reported. In the first one (De Robertis et al. 1982), ^{32}P labeled total RNA from HeLa cells was microinjected into the cytoplasm of *Xenopus* oocytes; snRNA's U_2, U_1, U_4, U_5, and U_6 were found to migrate and accumulate into the nucleus, 5S into the nucleolus while tRNA and 7S RNA remained in the cytoplasm. Human anti-Sm and anti-RNP antibodies were used to demonstrate that microinjected snRNA's bind to preexisting oocyte proteins accumulated in the cytoplasm prior to migrating into the nucleus. This was confirmed using enucleated oocytes in which the binding of microinjected snRNA's to snRNP proteins was also observed. This observation is well in keeping with the evolutionary conservation of both U snRNA's (Busch et al. 1982) and their associated proteins (M.R. Lerner and Steitz 1979; Wooley et al. 1982; Wieben and Pederson 1982). It also allowed the study of the distribution of snRNP during oogenesis and early development in *Xenopus* (Zeller et al. 1983). Small previtellogenic oocytes already contain a maximal amount of U_1 and U_2 snRNA's which then decreases to about 20% of that value in fully mature oocytes. At the same time, these fully mature oocytes, as well as embryos prior to gastrulation, were found to be depleted of snRNP's and to contain an excess of snRNP proteins stored in the cytoplasm. During late blastula, early gastrula or after microinjection of snRNA's in mature oocytes, snRNP's are assembled then migrate into the nucleus. It therefore seems that the accumulation of snRNP proteins is not coupled to the snRNA synthesis. In the same vein, *Drosophila* snRNA's microinjection into axolotl oocytes were found to migrate to the nucleus where they are protected from degradation (Saluz et al. 1983). More recently, Chandrasekharappa et al. (1983) using KB cells and Madore et al. (1984a) using HeLa cells, have confirmed the cytoplasmic assembly of snRNP's and demonstrated that prior to processing, U_1 snRNA precursor molecules are assembled into complexes containing the same antigen as nuclear U_1 snRNP.

Another reconstitution experiment was an in vitro assembly of snRNP's (Wieben et al. 1983a,b). HeLa cells poly $(A)^+$ mRNA was translated in a reticulocyte lysate and a mixture of snRNA's was added to the translation products. The assembly was monitored by the appearance of structures recognizable by anti-RNP or anti-Sm. With anti-Sm, it was shown that the same proteins than those which are selected in nuclear extracts are also selected in the translated products. However, when the assembly was monitored using U_1 snRNA and anti-RNP antibody, the resulting U_1 snRNA-protein complex was incomplete and contained only three newly translated proteins: one with a mol. wt. of 32 k, possibly corresponding to the U_1 specific protein previously identified by Hinterberger et al. (1983) and a closely spaced doublet of 27 k and 26 k. This finding has been extended by Fisher et al. (1984 and pers. comm.) who demonstrated, using pulse-chase experiments, that U_1 snRNP assembly is a two-step process involving prior assembly of a 6S core particle, comprising the small snRNP proteins, which then interacts with the U_1 snRNA and at least two (A and C) U_1 specific snRNP proteins.

Coherent with these findings is the suggestion that the assembly of a core structure characterized by the extremely tenacious binding of the set of small proteins to a common domain of the U snRNA's (Brunel et al. 1981; Liautard et al. 1982), could be a crucial step towards the reconstitution of snRNP's. Mattaj and De Robertis (1985), demonstrated that there is neither assembly of snRNP's nor transport into the nucleus, when an U_2 like RNA sequence, in which ten nucleotides surrounding the $A(U)_nG$ region (see B.VI below) have been replaced by fourteen other nucleotides, is injected in *Xenopus* oocytes.

VI. The Structural Organization of snRNP's

Whatever the function of any of the individual snRNP's turns out to be, a mandatory step towards unraveling the detailed mechanism of this function will be provided by the knowledge of the intimate structure of these particles, i.e., the way in which their constitutive elements interact with each other as well as with other eventual partners.

In an attempt to determine the conformation of U_1 snRNA in its RNP form, Epstein et al. (1981) submitted a nuclear extract from Novikoff hepatoma ascites cells to T_1 ribonuclease, then the digested RNPs were picked out of by anti-Sm antibodies. Comparison of the sites in U_1 snRNA which are accessible to cleavage either free or in RNP form revealed no significant differences, suggesting that no gross alteration of U_1 had taken place as a result of its packaging with proteins. The 5' terminal 107 nucleotides and most of the 3' end of U_1 RNA were resistant to cleavage and therefore assumed to be protected by proteins.

Another approach was used in our laboratory to locate on each snRNA the sites where proteins interact so strongly as to make the core particles resistant to the combination of CsCl and sarcosyl (Brunel et al. 1981; Sri-Widada et al. 1982). The digestion of core U_1 snRNP with up to 10,000 units ml^{-1} of micrococcal nuclease produced a main 25 nucleotides long resistant fragment spanning from nucleotide 119 to nucleotide 143 as proved by sequence analysis. When the same digestion protocol was applied to the mixture of "core snRNPs" containing all snRNA species, a more complex pattern of resistant fragments in the 15 to 35 nucleotides range was obtained. Fragments derived from all but U_6 snRNA's could be unambiguously identified by sequence analysis in the complex pattern of protected fragments. U_2, U_4, and U_5 protected regions were respectively from residues 94 to 115, 111 to 3'OH, and 83 to 3'OH (Liautard et al. 1982). All protected fragments had two striking features in common. Firstly, when matched to the most recently published models of secondary structure, they all belong to similar domains made up of a single-stranded region flanked on both sides by double-stranded stems, the 3' one extending to the terminus except for U_2 (Fig. 1). Secondly, all these fragments contained the sequence $A(U)_nG$ with $n \geqslant 3$ which has been conserved during evolution in *Drosophila* (Mount and Steitz 1981), man, rat, chicken (Branlant et al. 1980), dinoflagellate (Liu et al. 1984), and in plant cells (Krol et al. 1984. We propose that this domain with both its primary and secondary structure attributes is necessary for the recognition of core proteins and the stability of the structure. Moreover a recent observation by Mattaj and De Robertis (see above) shows that the $A(U)_nG$ sequence is necessary for cytoplasmic assembly of snRNP's and transport to the nucleus.

The core snRNP's which were used in the above protection experiments had a simpler protein composition than those purified either by more conventional biochemical methods or by immunoprecipitation (see Sect. B.V.5 above). We have now submitted to extensive digestion with micrococcal nuclease (5000 U ml^{-1}) a preparation of U_1 snRNP obtained by the procedure of Hinterberger et al. (1983). We found (Reveillaud et al. 1984) that most U_1 RNA sequese is protected except for the ten 5' terminal nucleotides presumably involved in splicing (see below in B.VIII.A) and a short sequence between nucleotides 102 and 108. Another region spanning nucleotides 60 to 79 is only weakly protected (Fig. 1). Mg^{2+} which is present in high concentration (15 mM) in this particle influences the binding of high mol. wt. polypeptides to U_1 snRNA. Indeed, if the particle is dialyzed to low Mg^{2+} they exhibits the same resistant structure as the U_1 core itself, the U_1 snRNP released from hnRNP at 0.1 M salt and 1.5 mM Mg^{2+} (Liautard et al. 1982) or the U_1 snRNP still interacting with hnRNP (Sri-Widada et al. 1983 and below in Sect. B.VII.A). As the dialyzed particle again becomes resistant when Mg^{2+} is adjusted back 15 mM, we suggested that U_1 snRNP undergoes a profound structural modification located in the 5' domain of the RNA molecule. Recently, Lin and Pederson (1984) have confirmed our findings on the structural organization of U_1 snRNP at high

Mg^{2+} except for some protection of the 5' terminus which in any case is nothing but marginal. As a matter of fact, a recent observation by Rinke et al. (1984) shows that 9 nucleotides at the 5' terminus of the intact U_1 snRNP are able to form stable DNA-RNA hybrids and are not protected by snRNP proteins.

Nicely coherent with the above findings is the evidence obtained by Agris et al. (1984) that the binding site for the high mol.wt. proteins containing the RNP antigenic determinant spans nucleotide 12 - 102 of the U_1 snRNA, i.e., does not overlap with either the 3' domain which interacts with the core proteins nor the 5' end which therefore remains available for interaction with splice junctions.

In our study on the ribonucleoprotein organization of U snRNP's (Reveillaud et al. 1984), we also investigated on the structure of U_2 snRNP using a partially purified particle obtained at high Mg^{2+} by the procedure of Hinterberger et al. (1983). Whatever the Mg^{2+} concentration, the U_2 snRNA sequence remains exposed in its 5' part up to nucleotide 92, in agreement with our previous result (Sri-Widada et al. 1983 and below in Sect. B.VII.A) that a part of this region is protected by hnRNP proteins when U_2 snRNP is interacting with hnRNP. However, Mg^{2+} also influences the structure of U_2 snRNP but only on the 3' side. At high Mg^{2+}, the hairpin III which is a characteristic feature of U_2 snRNA (Branlant et al. 1982) becomes partially protected (Fig. 1). Considering that U_2 snRNP seems to contain a specific protein (Kinlaw et al. 1983; Mimori et al. 1984), we suggested (Reveillaud et al. 1984) that this protein might interact in this region and participate to an actually unknown U_2 snRNP specific function. As a matter of fact, Mattaj and De Robertis (1985) have confirmed this localization using an anti-U_2 RNP serum.

In our study on the isolated U_5 snRNP (Lelay-Taha et al., submitted) we found this particle to be extremely stable. At high Mg^{2+}, only 6 supplementary nucleotides extending the core fragment become protected (Fig. 1). This suggest that all U_5 snRNP proteins may be located in the 3' domain of the RNA molecule.

As indicated above, no fragments derived from U_6 have been detected in our protection experiments on core snRNP's. The $A(U)_nG$ sequence, which appears as essential in the binding site for the common 9 - 14 k protein, is missing, suggesting a specific ribonucleoprotein organization for U_6 snRNA. Recent reports have confirmed and extended this last point. In their attempt at isolating native U snRNP's (see Sect. B.V.4), Bringmann et al. (1983a,b) observed that anti-2-2-7 trimethyl-guanosine antibodies precipitate all but U_6 deproteinized U snRNA's, whereas U_6 is coprecipitated when a cell extract containing snRNP's is reacted with antibodies. If this result highly suggests that U_6 interacts with some other snRNP, the crucial finding came from Hashimoto and Steitz (1984) and Bringmann et al. (1984) who demonstrated that U_4 and U_6 snRNA's coexist in a single snRNP (see Fig. 1, the diagram of the U_4 - U_6 particle).

VII. The Association of snRNP's with Rapidly Sedimenting Nuclear Structures

a) Biochemical Approaches

Early studies (Deimel et al. 1977; Guimont-Ducamp et al. 1977; Northemann et al. 1977; Gallinaro and Jacob 1979; Maxwell et al. 1981) had demonstrated the direct association of snRNA's with hnRNP which are assumed to be the site of post-transcriptional modifications of pre-mRNA. Quantitatively, it has been estimated that at least 40% of the snRNA's are associated with hnRNP in an average stoichiometry of one snRNA molecule for 2500 nucleotides of hnRNA (Gallinaro and Jacob 1977). Moreover, it has been shown that snRNA's interact with hnRNP's in the form of snRNP's recognizable by anti-Sm and anti-RNP antibodies (M.R. Lerner and Steitz 1979; Brunel et al. 1981; Barque et al. 1981). snRNA's have also been described as components of the nuclear skeleton (Zieve and Penman 1976; Miller et al. 1978; Vogelstein and Hunt 1982), which is formed by a protein matrix to which hnRNA and snRNA's are attached (Miller et al. 1978; Long and Schrier 1983; Long and Ochs 1983).

The nature of the association between low mol. wt. RNA's and messenger-type RNA's has been the subject of intensive investigation. Jelinek and Leinwand (1978) described the occurrence of a group of several species of low mol. wt. RNA (90 - 100 residues) which were hydrogen-bonded to both nuclear and cytoplasmic RNA, but these molecules are not U snRNA's but rather 4.5S RNA's. Flytzanis et al. (1978) have obtained experimental evidence in support of base-pairing interactions between snRNA and hnRNA on the basis of their dissociation at high formamide concentration, their reassociation under renaturation conditions, the resistance of their complexes to ribonuclease T1 and finally the isolation of approximately 10 - 25 base pairs long hybrids. Among these snRNA's, the two larger ones referred to as a and b but probably corresponding to U_1 and U_2, were the most tightly bound. Such an interaction was also observed by Maxwell et al. (1980). More recent reports provided evidence, based on in vivo cross-linking experiments with psoralen, that U_1 (Calvet and Pederson 1981) and U_2 snRNA (Calvet et al. 1982), are actually base-paired to hnRNA within the cell.

Moreover, the fact that snRNA's and especially U_1, are released from hnRNP by salt treatment (Gallinaro and Jacob 1979; Zieve and Penman 1981) strongly suggests that the association can also be mediated by proteins. Gallinaro and Jacob (1981) have reported detailed analyses of the localization of snRNA's within the two classes of rat brain hnRNP: the 30 - 50S monoparticles and the so-called heterogenous complexes (Gallinaro-Matringe et al. 1975; Stevenin et al. 1977; Gattoni et al. 1978). The main observation was that the different snRNA species are present in hnRNP of various sizes, but that U_1 is preferentially found associated with monoparticles. In addition, the authors have observed that a salt treatment, which preferentially dissociates monoparticles, induces the release of 50% of snRNA's and preferentially that of U_1 snRNA. When the salt treatment was completed by a digestion by proteinase K, only a small fraction of the snRNA's (10%) is further released, in agreement with the above observations that snRNA's can be directly associated with hnRNA. Knowing that snRNA's are already assembled into snRNP's in the cytoplasm (De Robertis et al. 1982), it seems likely that the two populations of snRNA's described above are both in snRNP form but are bound to hnRNP's through different types of interactions. That half which is released by high salt would be bound through protein-protein interactions, while the remaining half would be accounted for by snRNA's base-paired to hnRNA. In this context, Setyono and Pederson (1984 and pers. comm.) recently showed by in vivo cross-linking experiments that U_1 snRNA is in snRNP form when base-paired with hnRNA and that the A snRNP protein and a 42 k component from hnRNP bind at or near the 5' end of the snRNA.

In a work from our laboratory (Sri-Widada et al. 1983), we submitted total hnRNP from HeLa cells to extensive micrococcal nuclease digestion in order to ask whether additional snRNA sequences would be protected as a result of interaction with hnRNP constituents. In addition to the sequences protected by the core proteins, we found that U_2 and U_5 snRNA's exhibited highly protected domains located in their 5' region (Fig. 1). This suggests that the corresponding snRNP's tightly interact with hnRNP. This protection remains unchanged whether snRNP's are associated with monoparticles or higher size heterogeneous complexes both characterized by their protein composition (Gallinaro-Matringe et al. 1975; Stevenin et al. 1977; Gattoni et al. 1978). This indicates that the protection is not random and could depend on specific proteins. Interestingly, the U_5 sequence protected by hnRNP proteins is highly evolutionarily conserved (Branlant et al. 1983). No additional fragments derived from U_1 were found even a clone of U_1 DNA was used to select U_1 sequences. Thus U_1 snRNP is very accessible within hnRNP confirming previous observations (Gallinaro and Jacob 1979; Zieve and Penman 1981; Sri-Widada et al. 1982) on its easy extractability by salt treatment. A similar conclusion was reached by Branlant et al. (pers. comm.) by showing a great accessibility to dimethyl sulfate of U_1 snRNA within hnRNP.

Recently, Sass and Pederson (1984) have studied by immunofluorescence the location of snRNP's in the polytene chromosomes from *chironomus tentans* salivary glands. They found that a monoclonal anti-Sm antibody reacted with all sites of transcription (puffs and Balbiani rings). Furthermore, specific antibodies respectively directed against U_1 and U_2 snRNP's had identical patterns of localization and transcriptional dependence especially in Balbiani ring 2 known to contain only one mRNA transcription unit. This suggests a functional association not only of U_1 but also of U_2 snRNP to premRNP structures.

b) Electron Microscopic Approaches

Reports have been published (Daskal et al. 1980; Daskal 1981) that U_6 snRNA is localized in perichromatin granules which are most frequently found at the periphery of condensed chromatin and presumably associated with perichromatin fibrils (Monneron

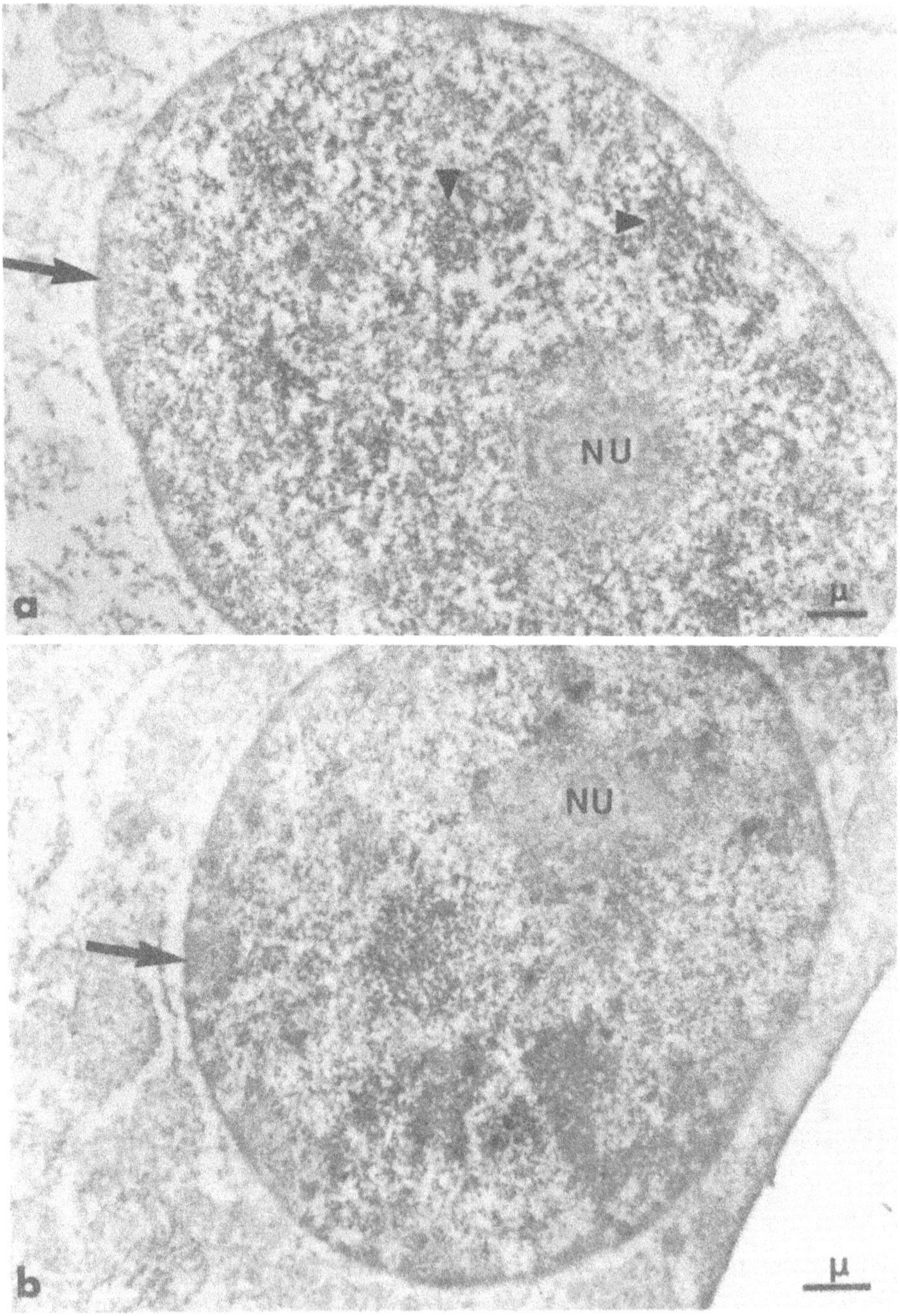

Fig. 2. a) Isolated rat liver cell grown in monolayer culture for 24 h were treated with anti-Sm Fab fragments purified by DNA-cellulose chromatography and revealed by peroxydase-labeled anti-human serum. The interchromatin RNP network appears stained *(arrowheads)* while the nucleolus (NU) and the dense chromatin *(arrow)* are not. (x 9000)

and Bernhard 1969). This indicates that U_6 snRNA may be released in vivo from the $U_4 - U_6$ snRNP (Hashimoto and Steitz 1984; Bringmann et al. 1984) and be involved in free form in a nuclear function different from those of other snRNP's. In agreement with this last point is our earlier observation that the snRNP form of U_6 is unstable (Brunel et al. 1981) and the fact that about 50% of U_6 snRNA are released from antibody-bound snRNP's by exposure to 37°C (Hashimoto and Steitz 1984).

The nuclear distribution of Sm and RNP antigens have been studied recently by immunocytochemistry using purified Fab fragments from the corresponding antibodies of human origin (Puvion et al. 1984). This distribution was found to exactly overlap the ribonucleoprotein pattern of nuclei from isolated rat hepatocytes, i.e., the perichromatin fibrils and granules as well as the interchromatin granules (Fig. 2). When this pattern is altered by treatment of the cells with DRB (5,6 dichloro-1-βD-ribofuranosyl benzimidazole), a drug known to inhibit the transcription of premessenger RNA (Tamm and Kikuchi 1979) an important accumulation of interchromatin granules ensues which correlates with an intense labeling by antibodies. Similar results have been described by Fakan et al. 1984.

VIII. Some Hypotheses About the Function of U snRNP's

Of all six U snRNA's, only U_1, U_2, and U_3 and recently U_4 have been the subject of elaborate models aimed at explaining their function. For lack of identified enzymatic activities among their proteins, snRNP's appear in all these models as recognition factors could mediating interaction between primary sequences and other components. Other models involving other species of low mol. wt. RNA, like the adenovirus-coded VA RNA's and a 4.5S, have also been proposed. The case of VA RNA will be covered along with that of U_1 to which it is formally very close.

a) U_1

The very first model involving a low molecular RNA in splicing of premessenger was proposed by Murray and Holliday (1979). It relied on a striking complementarity between VA_1 and the exon side of the splice junction between leader 3 and the remaining of hexon mRNA. No experimental evidence has yet been found in support of this model. Indeed, it is puzzling to note the nearly complete absence of VA in hnRNP from adenovirus-infected HeLa cells (Gallinaro et al. 1980, 1981), the absence of VA RNA in nascent hnRNP's containing adenovirus pre-mRNA (Blanchard 1980) and the lack of inhibition of splicing by anti-La antibody which

◄ Fig. 2. b) Isolated rat liver cell grown in monolayer culture for 24 h and treated with DRB for 2 h before collection. The immunocytochemical procedure is the same as in *a*. The DRB-induced clumps of interchromatin granules (IG) are stained. The altered nucleolus (NU) and the dense chromatin *(arrow)* are not stained as in *a*. (x 9000). (Puvion et al. 1984)

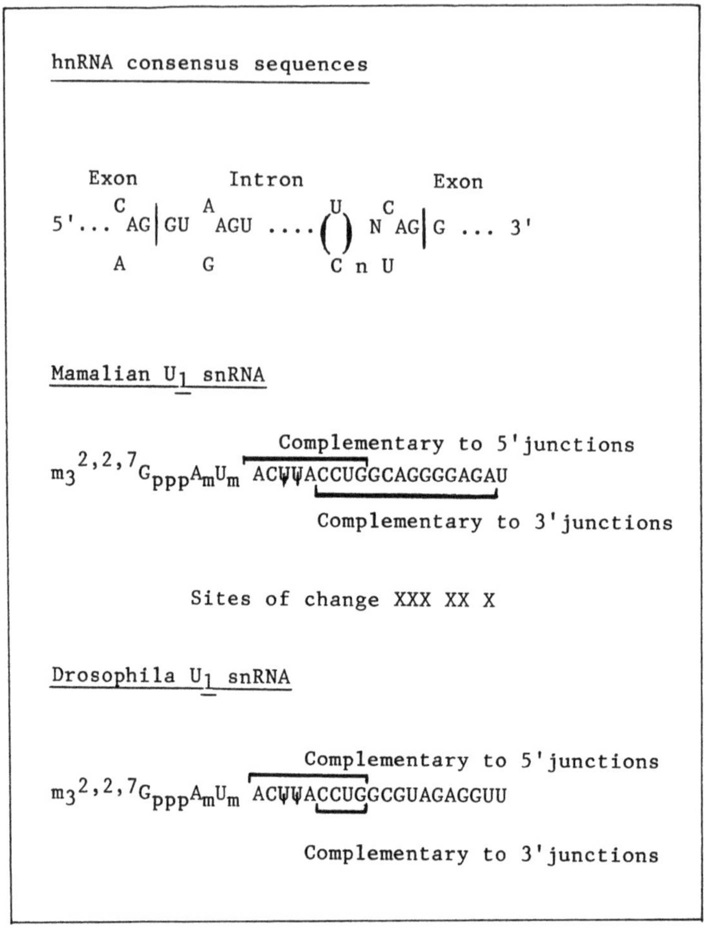

Fig. 3. U_1 snRNA and splicing (Steitz et al. 1983)

recognizes VA RNA's (Yang et al. 1981). In fact, a role for VA_1 RNA in the regulation of adenovirus 5 mRNA translation has recently been proposed and well documented (Thimmappaya et al. 1982) (see C.I.3.a) below).

Very interesting models in the same vein have been proposed for U_1 snRNA. They all stemmed from the observation that the 5' terminal nucleotides of U_1 shows complementarity with an hnRNA consensus sequence derived from the examination of numerous intron-exon boundaries (M.R. Lerner et al. 1980; Rogers and Wall 1980). The original proposal was a cross-over model (Fig. 3) in which nucleotides 4 to 21 at the 5' end of U_1 could base-pair with both the acceptor and donor sides of introns (Breathnach and Chambon 1981; P. Sharp 1981; Mount 1982). By examining the sequence of U_1 snRNA from drosophila, Mount and Steitz (1981) noted that only three out of the eight nucleotides between 14 and 21 in U_1 are conserved from insect to human and that this region is not exclusively made of purines in *Drosophila* as it is in man. Moreover, this 14 - 21 region appears to be involved in secondary structure interactions within U_1. A revised model was then proposed (Mount and Steitz 1981, Steitz et al. 1983) in which U_1 snRNA, presumably in its RNP form, recognizes only the 5' splice site (Fig. 3). Only the 5' terminus of the U_1 snRNA $m_3\,G_{ppp}A_mU_m$ACUUACCUG which is unchanged from insects to vertebrates would therefore be involved in this recognition. Strong biochemical evidence in support of this revised model came from the demonstration by Mount et al. (1983) that a partially purified U_1 snRNP, free of RNAse activity (Hinterberger et al. 1983), bind specifically to an in vitro artificial transcript which contains the sequences corresponding to the small intron and part of the flanking exons of the major mouse beta globin gene. Very interesting is the observation that snRNP proteins contribute to the recognition since naked U_1 snRNA does not compete with U_1 snRNP for binding, and a proteinase K digestion of U_1 snRNP abolishes the interaction by 80%.

In addition to the evolutionary conservation of U_1 snRNA sequences, and even more so at the 5' terminus, which can be paralleled with the conservation of the splice sites, many other data are consistent with a role for U_1 snRNP in the recognition of the splice junctions in hnRNA. They include: the abundance of U_1 snRNP in the cell, particularly in rapidly dividing cells and its association with hnRNP in which splicing takes place; the fact that U_1*, a shortened form of U_1 snRNA which lacks the 5' cap structure and the first six nucleotides, no longer binds to hnRNP's (M.R. Lerner et al. 1980); the observation that some U_1 snRNA molecules are base paired to hnRNA in hnRNP (Flytzanis et al. 1978) and can be cross-linked in vivo (Calvet and Pederson 1981), the possible base-pairing of free U_1 snRNA with a single-stranded DNA fragment mimicking the intron borders at the splice junction (Lazar et al. 1982), the observation that autoimmune antibodies inhibit splicing of adenoviral molecules in isolated nuclei (Yang et al. 1981); the important reduction of VP-1 protein and large T-antigen but not small t-antigen synthesis (Fradin et al. 1984) after co-injection in *Xenopus* oocytes of SV40 DNA and serum from SLE patients or the demonstration that splicing of *Xenopus laevis* ribosomal protein mRNA's is inhibited in vivo by antisera to snRNP's (Bozzoni et al. 1984); the observation that in vitro splicing of the Adenovirus 2 major late premRNA is stimulated by partially purified snRNP's (Hernandez and Keller 1983). The crucial demonstrations finally came along this last line from: (a) The use of an in vitro splicing system which is inhibited by monoclonal anti-Sm, patient anti-Sm or anti-RNP antibodies but not by an SLE serum precipitating U_2 snRNP (Padgett et al. 1983). (b) The finding that upon removal of the first eight nucleotides of the U_1 snRNA in the particles by site-directed hydrolysis with ribonuclease H in the presence of a synthetic complementary oligodeoxynucleotide, splicing is completely abolished (Kramer et al. 1984). However, on the basis of the results of Fradin et al. (1984) showing that Sm antibody does not affect the SV40 small t-antigen synthesis, it seems that different pre-mRNA's have dissimilar interactions with snRNP's in the splicing reaction.

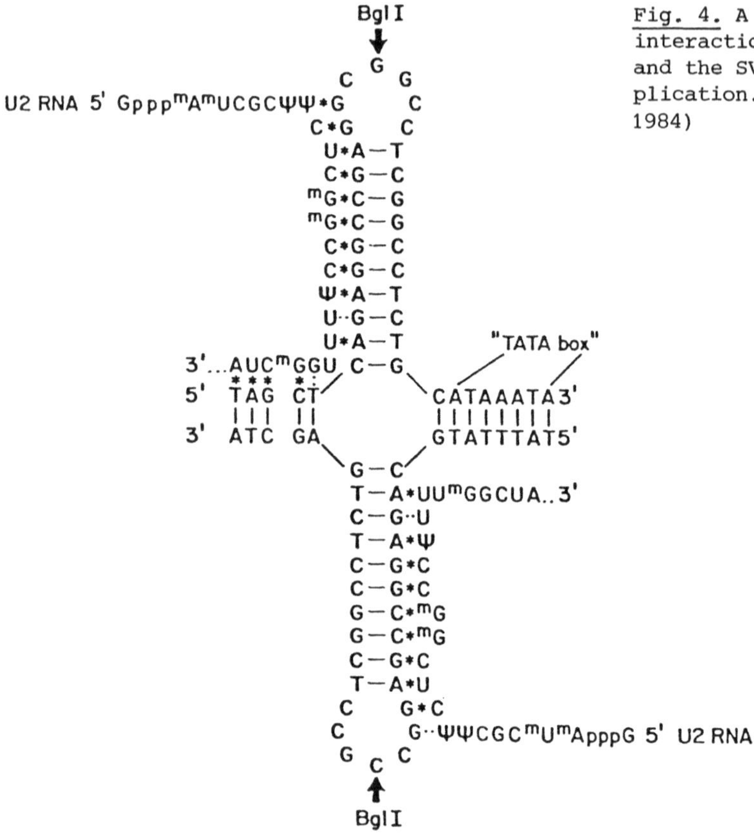

Fig. 4. A model of potential interaction between U_2 snRNA and the SV40 origin of replication. (Savouret et al. 1984)

b) U_2

Other hypothetic models have been proposed that involve interactions between U snRNA's or other small RNA's and mRNA precursors (Harada et al. 1980; Ohshima et al. 1981a). One was the suggestion that U_2 snRNA could also be involved in the splicing of certain mRNA precursors by hybridizing to exonic regions across the splice junction (Ohshima et al. 1981a). This, however, is not supported by the secondary structure model for U_2 snRNA. Indeed, the region assumed to engage in such an interaction comprises a 21 nucleotides long sequence spanning nucleotides 120 - 140 in the model of Branlant et al. (1982) and among these, only 8 nucleotides are not involved in secondary structure interactions. Another similar interaction was proposed by Harada et al. (1980) between U_6 snRNA and sequences of the splice junctions of some mRNA precursors.

An interesting observation was made recently (Savouret et al. 1984) showing that a fraction of cellular U snRNP's are firmly associated with the viral chromosome in CV-1 cells infected by SV 40. If the association of U_1 snRNA with the minichromosome can be explained by its requirement for viral transcript splicing according to the model of M.R. Lerner et al. (1980), the presence of other snRNA's especially U_2, was unexpected. Examination of sequences revealed a striking complementarity between the 5' end of U_2 snRNA and the DNA sequences involved in the cruciform structure at the replication origin of SV 40 (Fig. 4). U_2 complementarity spans 17 nucleotides on the right strand for replication with only one mismatch. If any, this interaction could be the mechanism by which the DNA structure is disrupted or modified allowing SV 40 to switch from a replication to a transcription mode.

c) U_3

It has been known for a long time that U_3 is hydrogen-bonded to 45 and 32S pre-ribosomal RNA when total RNA is extracted at low temperature (Prestayko et al. 1970). Its potential role in the processing of 32S into 5.8S and 28S rRNA has been proposed (Prestayko et al. 1970; Zieve and Penman 1976). Two models of interaction between U_3 and the ribosomal RNA precursors are currently available. The first one (Crouch et al. 1983) proposes that U_3, along with its cognate proteins, acts as an RNA isomerase by interacting with both the 3' terminus of 5.8S RNA and the second internal transcribed spacer (ITS-2) to induce a configuration of the precursor adequate for splicing. The second model (Bachellerie et al. 1983) is based on the possibilities of base-pairing interactions between U_3 and the 32S precursor, especially in a region of the ITS-2 which extends precisely to the cleavage site generating the 3' terminus of the 5.6S RNA. In this model, the ITS-2 is considered as formally equivalent to an intron. Recently, Reedy et al. (pers. comm.) have demonstrated that the interaction of U_3 snRNP with the pre-ribosomal RNP particles is for the largest part mediated by protein-protein interactions. Only 13% of the U_3 snRNA seems to be hydrogen-bonded to pre-ribosomal RNA.

d) U_4

In recent work describing a cell-free transcription reaction that generates polyadenylated RNA at the L_3 site of the major late precursor of Adenovirus, Moore and Sharp (1984) observed that the reaction is inhibited by sera of Sm, RNP, and La specifities, suggesting the involvement of some of snRNP's in the polyadenylation process. Among these, U_4 snRNP seems to be the best candidate and two models have emerged recently, both assuming that U_4 mediates polyadenylation in a similar manner as U_1 snRNP mediates splicing. In the first one, Berget (1984) proposed pairing of U_4 RNA both to the AAUAA recognition signal and to the CAYUG which is adjacent to the site of polyadenylation of numerous sequenced RNA's. In the second model (Sadofsky and Alwine, pers. comm.), the sequence 3'-CCUUUUCAAAAGUUAAUC-5' which appears in a single-stranded loop of U_4 snRNA recognizes a consensus sequence located within 20 - 60 nucleotides downstream from the AAUAAA signal.

On account of the recent demonstrations that U_4 and U_6 snRNA's coexist in the same particle (Hashimoto and Steitz 1984; Bringmann et al. 1984), a modulating role of U_6 snRNA in the U_4 snRNP activity should be considered.

C. Other Small RNA's Precipitable with Autoimmune Antibodies

As an extension of studies on U snRNP's which are precipitable by anti-Sm and anti-RNP antibodies, it has been shown in recent years that other auto-immune antibodies (especially anti-Ro or SS-A and anti-La or SS-B, both present in sera from patients with Sjögren's syndrome or SLE), immunoprecipitate RNP structures containing a wide spectrum of small mol. wt. RNA's.

Anti-Ro precipitates stable small cytoplasmic RNA's (Y scRNA's) which are present in low abundance and whose number varies from one mammalian species to another. Anti-La precipitates numerous RNA species like the nuclear 4.5S and $4.5S_1$, tRNA precursors, 7.2 RNA, pre-5S RNA as well as viral RNA's (adenovirus-encoded VA_I and VA_{II}, Epstein-Barr virus encoded EBER 1 and 2) and also Ro RNA's which can be considered as a subclass of La RNA's.

Since all these RNA's are either assumed or demonstrated to be transcribed by RNA-polymerase III, the possibility that La antigen may function in the transcription or maturation of all transcripts made by this polymerase has recently been largely investigated. An interesting finding is that several RNA-polymerase III transcripts (pre-5S RNA, pre-tRNA and 7.2 RNA) transiently bind La antigen before being processed and incorporated into RNP's recognizable by other autoimmune antibodies. Along the same line, it was recently observed that anti-La antibodies immunoprecipitate in vitro transcripts of Alu-family repeats by RNA-polymerase III thus demonstrating their interaction with La antigen (Shen and Maniatis 1982).

In two cases, RNA's which are not tanscribed by RNA polymerase III have been found to be transiently associated with La antigen. One is the leader RNA of Vesicular Stomatitis Virus which can be immunoprecipitated from infected BHK cell extracts 4 h after infection (Kurilla and Keene 1983). Surprisingly, the second RNA is the U_1 snRNA or its precursor forms (Madore et al. 1984b). In both cases, it seems that the presence of uridylate residues at the 3' end may be responsible for interaction with La antigen.

I. La Antigen Binds RNA-Polymerase III Transcripts

We will distinguish here the RNA's which don't seem to be processed after transcription and those which lose their La antigen as their are processed. The first ones include viral RNA's (VA_I, VA_{II}, EBER 1, EBER 2) as well as the 4.5S RNA's and cytoplasmic Y RNA's which are also recognized by anti-Ro. Among the second ones are tRNA precursors, and pre-5S RNA. Although not being processed, nucleolar 7.2 RNA has also been found transiently associated with La antigen before being incorporated into another type of antigenic RNA (Hashimoto and Steitz 1983). The case of 7S RNA, which is partially complementary to Alu sequences (Weiner 1980) and recently shown to be required for protein translocation across membranes (Walter and Blobel 1982) will be dealt with below (Sect.D.III).

1. 4.5S RNA's

4.5S RNA's from Novikoff hepatoma cells consist of three distinct species which can be separated by column chromatography (Ro-Choi et al. 1970) and then by two-dimensional gel electrophoresis (Reddy et al. 1974b). One of these three, the 4.5S was the first snRNA to be sequenced (Ro-Choi et al. 1972). Homologies have been noted (Gojobory and Nei 1981) between 4.5S RNA and the flanking regions of eukaryotic genes (Hogness box and Proudfoot-Brownly box), suggesting that it might be involved in the recognition of some transcription signals. This hypothesis has received no experimental support yet.

Jelinek and Leinwand (1978) have reported a group of 90 - 100 nucleotides long RNA's which were isolated by melting them away from poly(A)$^+$ nuclear and cytoplasmic RNA from Chinese hamster ovary (CHO) cells, suggesting their possible implication in the production and/or utilization of mRNA. The nuclear function of these RNA's is supported by the finding that nuclear poly(A)$^+$ molecules contain them in a molar ratio about seven times higher than in cytoplasmic poly(A)$^+$ RNA. Interestingly, the 4.5S$_1$ RNA from CHO cells was found to be partially homologous to the highly repetitive interspersed sequences called Alu family (Haynes and Jelinek 1981) and to the sequence that comprises one half of the perfect inverted repeat at or near the origin of DNA replication of papovaviruses (Jelinek et al. 1980).

4.5S$_1$ RNA with similar properties of hydrogen-bonded to poly(A)$^+$ molecules have been described in mouse, rat, and hamster cells (Harada et al. 1979) and sequenced in mouse and hamster cells (Harada and Kato 1980) in which there is only one base substitution. This RNA is different from the 4.5S$_1$ RNA from Novikoff hepatoma (Ro-Choi et al. 1972) and was not found in other vertebrates like man, monkey, cat, mink, rabbit, guinea pig, or chicken. It contains two sets of repeated sequences, a central purine-rich sequence and a region complementary to some splice junctions as in the case of U$_1$ snRNA. Since this RNA is associated with nuclear and cytoplasmic poly(A)$^+$ RNA, it was suggested that it could be involved not only in splicing but also in transportation of mature mRNA.

In order to determine whether human cells contain RNA with characteristics analogous to 4.5S RNA from mouse or hamster cells, Kato et al. (1982) have analyzed low mol. wt. RNA's from HeLa cells. An 83 - 84 nucleotides long species containing analogous features, namely a 5'pppA and a 3' oligo(U) stretch, was found. It seems that this RNA may be identical to the hY$_5$ RNA immunoprecipitated by anti-Ro antibodies (see below) since these two RNA's have the same T$_1$ fingerprint, the same 5' terminus, and the same cellular localization. This human 4.5S RNA has no homology with the Alu family but is appreciably reminiscent of the 1.9 kb Hind III family described by Manuelidis (1982).

2. Ro RNA's

Due to their low abundance in the cell (1 to 5×10^5 copies per cell, Hendrick et al. 1981), Ro RNA's have only been detected by antibodies. From 80 to 110 nucleotides in size, their number varies among mammalian species: two in mouse, mY_1 and mY_2, three in rat, rY_1 to rY_3, and five in man, hY_1 to hY_5. First described as antigenically distinct from La RNA's (E.A. Lerner et al. 1981), the presence of hY_5 in the spectrum of RNA's precipitated by anti-La suggested that Ro RNA's might be also associated with La antigen (Hendrick et al. 1981). This was confirmed by reconstitution experiments in which Ro particles isolated by immunoadsorption to an anti-Ro column were used as a source of proteins for reassociation with total cellular RNA in a form recognizable by anti-La (Hendrick et al. 1981). Ro RNA's have 5' triphosphate termini and lack modified nucleotides (Hendrick et al. 1981). Francoeur and Mathews (1982) have found that two major proteins of 90 and 94 k are precipitated by anti-Ro sera. However, Wolin and Steitz (1984) show that an antigenic polypeptide of 60 k is the major protein component of Ro RNP's.

A genomic clone containing sequences complementary to human Ro RNA's has recently been analyzed (Wolin and Steitz 1983). It contains true coding regions for hY_1 and hY_3 RNA's suggesting a genomic organization of Ro RNA's genes quite different from that generally observed for U snRNA's (see B.IV). Genes for hY_1 and hY_3 appear to be single-copy in the human genome and therefore are the first example of mammalian single-copy genes not encoding proteins. They appear to be transcribed by RNA polymerase III in HeLa cell-free extracts since their transcription is sensitive only to high concentrations of α-amanitin. When the DNA from this clone was used for hybrid-selecting RNAs, hY_1, hY_2, and hY_3 RNA's as well as two smaller ones called hY_3 and hY_3** were identified in anti-La immunoprecipitates. This finding is in agreement with the idea that Ro RNA's are a subclass of La RNA's. hY_2 was shown to be closely related to hY_1, differing only by its 5'pG terminus (instead of pppG) and the lack of two oligonucleotides at the 3' end. hY_3* and hY_3** were also selected in anti-La precipitates by this method but hY_3** did not react with anti-Ro antibodies suggesting that it lacks the binding sites for the Ro antigen. The same kind of experiment also showed that Ro RNA's are quantitatively contained in Ro scRNP's. Homologies were found between human and mouse Ro RNA's especially between hY_1 and mY_1 which therefore appear to be the most conserved during evolution. Hypothetical secondary structures for hY_1 and hY_3 were derived from this work (Wolin and Steitz 1983) and compared to that proposed for hY_5 by Kato et al. (1982). All these RNA's are characterized by the presence of a pyrimidine-rich internal loop and a stem in which the 3' and 5' terminal regions of the molecule are hydrogen-bonded together. This stem appears as a binding site for the Ro-antigen (Wolin and Steitz 1984).

Like the adenovirus-encoded VA RNA's which appear essential for the translation of late adenovirus mRNA (Thimmappaya et al. 1982), Ro RNA's might play a role in regulating the translation of certain mRNA's.

3. Viral RNA's

We will limit the discussion to small viral RNA's which have been found to bind the La antigen, namely VA_I and VA_{II} transcribed from the adenovirus genome in infected cells and two EBV encoded RNA's (EBER 1 and EBER 2). Other small viral RNA's associated with SV40 (Hutchinson et al. 1979; Alwine et al. 1980), standard and defective VSV particles (Rao and Huang 1979; Emerson et al. 1977; Schubert et al. 1978) or retrovirus (Bishop et al. 1970; Sawyer et al. 1974) have been also described. In this last case, however, the virus associated RNA is not viral encoded and corresponds to the host 4.5S recognized by anti-La antibodies (Harada and Ikawa 1979; Harada et al. 1979) or to the host 7S (Erikson et al. 1973; Larsen et al. 1973, 1974) (see Sect. D).

a) VA RNA's

In the viral genome there are two RNA's encoded by the r-strand near position 30 of the adenovirus type 2 physical map. They are separated by a short DNA sequence (Mathews and Petterson 1978). One codes for the 157 - 160 nucleotides long VA_I RNA, the other for a less abundant (about 40 : 1) 158 - 163 nucleotides long VA_{II} RNA. These VA RNA's are unique among adenovirus transcripts in being synthesized by RNA polymerase III (Söderlund et al. 1976; Weinman et al. 1976; Harris and Roeder 1978; Weill et al. 1979). They do not undergo post-transcriptional processing at their 5' termini (Price and Penman 1972; Wallace and Kates 1972; Söderlund et al. 1976; Weinmann et al. 1976) and the promoters for VA_I have been shown to be contained within the structural gene itself (Fowlkes and Shenk 1980; Guilfoyle and Weinmann 1981). Adenovirus-2 DNA sequences that span both VA_I and VA_{II} as well as a connecting spacer of 98 bp (Akusjarvi et al. 1980) have suggested that these two genes may have arisen by duplication of an ancestral gene with subsequent divergence. Secondary structures were derived either by nuclease digestion experiments in the case of VA_I (Monstein and Philipson 1981) or by computer for VA_I and VA_{II}.

A role for VA_I RNA has been proposed recently (Thimmappaya et al. 1982). Two adenovirus 5 variants have been constructed each of which fails to synthesize one of the VA species. The VA_I^+/VA_{II}^- mutant grows normally, but the VA_I^-/VA_{II}^+ grows more poorly than its wild-type parent. Cells infected with the VA_I^-/VA_{II}^+ mutant contained normal levels of both early and late viral mRNA's which appear to be normally capped, spliced, polyadenylated, localized in the cytoplasm of host cells and efficiently translated in vitro. However, the levels of both early and late polypeptides at late times after infection were reduced, demonstrating that VA_I RNA is required for efficient translation. The rate of nascent polypeptides elongation being the same in cells infected both with the wild-type virus or the VA_I/VA_{II}^+ mutant, attractive roles for the abundant VA_I RNP during initiation have been proposed. VA_I RNP might serve to attach mRNP's to the cytoskeleton where they are translated (Cervera et al. 1981). Alternatively, the VA_I particle may direct initiation factors to viral mRNA's competing with cellular RNP's, possibly the Ro scRNP's which serve a similar role in the synthesis of cellular proteins.

Both VA_I and VA_{II} RNA's associate with host proteins giving rise anti-La recognizable structures (E.A. Lerner et al. 1981; Rosa et al. 1981).

b) EBER RNA's

EBER 1 and EBER 2 RNA's have been shown to be synthesized in lymphoid cells transformed by Epstein-Barr virus (EBV) (E.A. Lerner et al. 1981). As pointed out by Rosa et al. (1981), striking similarities are exhibited by VA and EBER RNA's. EBER RNA's are direct products of the EBV genome and are transcribed by RNA polymerase III. Correlating the nucleotide sequence of the EBV genome encoding the two RNA's and RNA fingerprint data, have demonstrated that DNA sequences corresponding to EBER 1 and EBER 2 RNA's are contained within a 707 bp HaeIII fragment (Rosa et al. 1981) derived from the B95-8 strain of EBV (Dambaugh et al. 1980). EBER 1 and EBER 2 are respectively 166 - 167 and 172 +/- 1 nucleotide long RNA's and possess the same 5' terminal oligonucleotide pppAGGAC. They are separated by a 161 bp spacer and transcribed from the same DNA strand. In the region preceding the EBER 1 coding sequence, there are two 9 bp sequences which are repeated at the same place before the EBER 2 coding sequence. Therefore, EBER 1 and EBER 2 may have arisen by duplication of an ancestral gene. Within the coding regions, there are homologies at the 5' termini. Two of these regions are in locations analogous to intragenic RNA polymerase III transcription control region identified in adenovirus 2 VA_I RNA (Fowlkes and Shenk 1980). Hypothetical secondary structures models have been drawn (Rosa et al. 1981) resembling that of VA RNA's.

Just like VA RNA's, EBER RNA's associate with proteins to form an anti-La recognizable structure and particles can be assembled in vitro by mixing ^{32}P labeled RNA's, from EBV-transformed or adenovirus-infected cells with homologous as well as with mouse cell extracts, suggesting the conservation of La antigen across species (Rosa et al. 1981). Discussing their work, these authors have suggested a common function for VA and EBER RNA's. Like VA_I RNA, EBER RNA's might then function also at the level of translation in transformed cells, an hypothesis which is actually being tested (Thimmappaya et al. 1982).

4. Pre 5S RNA and Pre tRNA's

In the bulk of RNA's which are immunoprecipitated by anti-La antibodies, Rinke and Steitz (1982) have identified a 5S* species which migrates slightly lower than ribosomal 5S RNA. RNAse T_1 fingerprint analyses revealed that 5S RNA is a mixture of 5S-like molecules ending with $CUUU_{OH}$ (corresponding to one form of 5S RNA, the other ones being CUU_{OH}, $CUUUU_{OH}$ and $CUUUUU_{OH}$ in relative ratios of 1:2:0.5). Pulse-chase experiments made both in vivo and in isolated nuclei have demonstrated that 5S* is a precursor to mature 5S RNA and that, since the antigenicity is lost as mature 5S RNA molecules appear, La antigen binds preferentially to such a precursor. In vitro synthesized 5S transcripts are converted into mature size molecules by incubation in a cell extracts, indicating the presence of a processing activity that was not further characterized.

In the same vein, Rinke and Steitz (1982) have shown that La antigen binds tRNA precursors as expected by Hendrick et al. (1981) in view of the fact that many La RNA's from HeLa cells were hybridized to immobilized recombinant DNA carrying single genes for $tRNA^{Met}$ (human), $tRNA^{Glu}$ (rat), $tRNA^{Asp}$ (rat), $tRNA^{Gly}$ (rat) and $tRNA^{Asn}$ (human) and compared to total in vivo labeled RNA hybridized with the same clones. As expected, RNA's hybridized from the La immunoprecipitate were found to migrate more slowly than the corresponding mature tRNA's and the presence of multiple bands characteristic of tRNA's precursors undergoing processing was noted. By RNAse T_1 fingerprint analysis, it was established that the La RNA's are precursor forms to mature tRNA molecules. In vitro synthesized precursors of $tRNA^{Met}$ were found to be immunoprecipitated by anti-La antibodies, whereas products which have been processed into mature tRNA were found in the supernatant. Once processed, 5S RNA (E.A. Lerner et al. 1981) and at least one tRNA (Rosa et al. 1983) bind to protein(s) (see Sect. C.III) and form particles which are recognized by other antibodies. However, it seems that some tRNA's may not require the presence of proteins to be antigenic (Hardin et al. 1982; Epstein et al. 1982; Rosa et al. 1983).

5. Nucleolar 7-2 RNA

7-2 RNA is one of the four 7-8S RNA's (7-1, 7-2, 8A, and 8B) which have been reported to be nucleolar on grounds of biochemical fractionation experiments using Novikoff hepatoma and HeLa cells. Hashimoto and Steitz (1983), have observed that anti-La antibodies consistently immunoprecipitate HeLa and Friend erythroleukemia RNA's that migrate slightly slower than 7-2 RNA and which were shown to be transcribed by RNA polymerase III. Both RNA's showed essentially the same T_1 and pancreatic RNAse fingerprints and the only difference was at the 5' end; while 7-2 RNA is enriched in 5' terminal pG, the La immunoprecipitated 7-2 RNA has mostly pppG. On the basis of this finding, Hashimoto and Steitz (1983) have suggested a precursor-product relationship, although the remote possibility that the La 7-2 and 7-2 RNA's are differently processed transcipts from the same gene, or transcripts from two differently sized 7-2 RNA genes, cannot be ruled out.

In the same work, Hashimoto and Steitz (1983) have found that 7-2 RNA is immunoprecipitated by a serum from a patient with SLE called anti-Th (Hardin et al. 1982) and have observed that this serum always immunoprecipitates predominantly the faster migrating 7-2 RNA vs. the slower migrating form which is recognized predominantly by anti-La antibodies. Thus, like pre-5S RNA and pre-tRNA's, the nucleolar 7-2 RNA seems to be transiently associated with La antigen before being incorporated into another form of particles. However, some 7-2 RNA may be bound simultaneously to both Th and La antigens.

Interestingly, this anti-Th serum was found to immunoprecipitate Ro RNA's and a cytoplasmic 8-2 RNA newly identified by Reddy et al. (1983b), using a serum from patient with Scleroderma designated anti-To. Although this anti-To antibody also recognizes 7-2 RNA (Reddy et al. 1983), it remains to be elucidated whether these two antibodies recognize a common protein determinant and whether Ro scRNP's share with 7-2 and 8-2 RNP's a protein recognized by anti-Th.

Just before the up-cited works (Hashimoto and Steitz 1983; Reddy et al. 1983b) were published, Nakao et al. (1982) have found a serum from patients with polymyositis that recognize a ribonucleoprotein complex containing 7.5S RNA. This 7.5S RNA consists of two species having slightly different electrophoretic mobilities and was found to be located both in the nucleolus and the cytoplasm by immunofluorescence staining. It seems that the two 7.5S species might be identical to 8A and 8B RNA's described by Reddy et al. (1981b), but as emphasized by the authors the possibility remains that one of the 7-1 and 7-2 RNA's might correspond to 7.5S RNA or that 7.5S RNA might be a previously unidentified species.

II. Molecular Nature and Function of La Antigen

The finding that RNA polymerase III transcripts associate at least transiently with La antigen have suggested that the protein carrying the La determinant plays an important role in the biogenesis of RNA polymerase III products. Therefore efforts are actually being made in order to determine its molecular nature. Steitz and her colleagues (1983) now describe preliminary results along this line. Using the Western blot technology developed by Towbin et al. (1979), it appears that anti-La patients have antibodies reactive with a single polypeptide of about 50 k. This protein appears to be required for the synthesis of EBER 1 and EBER 2 RNA's, since cells extracts depleted in La-protein by exposure to Sepharose-linked antibody failed to transcribe EBER genes. In addition, anti-La antibodies were found to precipitate a vector containing cloned VA_I and VA_{II} genes after incubation in a transcriptionally active extract indicating that the protein carrying the La determinant interacts either with the gene itself or with the transcript still attached on the DNA. Therefore the La-protein might be considered as a general RNA polymerase III transcription factor, perhaps corresponding to one of many factors which have been previously described as being required for accurate transcription by RNA polymerase III (Segall et al. 1980; Honda and Roeder 1980; Pelham and Brown 1980; Engelke et al. 1980).

Francoeur and Mathews (1982) have described the La-antigen as being a unique phosphorylated protein of 45k, and proceeded to show that it is composed of at least 8 isoelectric species which are phosphorylated and react with anti-La antibodies (Francoeur and Mathews, pers. comm.). This protein assembles with VA RNA's in transcription systems programmed for VA RNA's synthesis, as well as by mixing nondepleted cell extracts and purified VA RNA's or altered forms of VA RNA's. The binding site for La-protein was found to comprise the 5' terminal 13 nucleotides and the 3' terminal 26 nucleotides, but it seems that the uridylate residues that generally terminate RNA-polymerase III transcripts may be of importance, since blocking of the 3' end eliminates La-antigen binding (Stefano 1984). Pizer et al. (1983) have observed that the level of phosphorylation of La-antigen was increased in adenovirus-infected KB cells.

In agreement with these findings on the nature of La-antigen are the work from Van Eekelen et al. (1982) in which a polypeptide with a mol. wt. of 50 k was found to be associated with adenovirus coded VA RNA after ultraviolet light-induced RNA-protein cross-linking and the work from Habets et al. (1983b) showing that the 50 k component is well conserved during evolution. At variance with the preliminary results reported by Steitz et al. (1983) on the requirement of La-antigen for EBER RNA's synthesis, Francoeur and Mathews (1982) have found that La-antigen is not required for VA RNA's synthesis since La-antigen depleted extracts were found to be competent for VA_I RNA synthesis and that the addition of antibodies failed to inhibit VA_I RNA synthesis in such a system. Thus, from this work, La-antigen, defined as a unique protein of about 50 k, does not appear to be a transcription factor.

The final demonstration along this line came with two recent reports from Hoch's group (Hoch and Billings 1984; Gottesfeld et al. 1984). They have developed a single two-step procedure (sonication of a whole cell extract which is passed directly through an immunoaffinity column) to isolate the La-associated RNP complex from rabbit or rat thymus acetone powder or from HeLa cells. In addition to a main 50 k component, this La-RNP was found to contain four bands among which was a 64 k polypeptide. They have first confirmed by immunoblotting that the La-antigen is indeed the 50 k polypeptide, then proceeded to show that in vitro transcription of several genes, like a cloned *Xenopus laevis* 5S RNA gene (oocyte type), a *Xenopus* $tRNA^{Met}$ gene and the adenovirus VA_I gene, can be inhibited by anti-La antibodies. However this inhibition could be restored both by the addition of an S100 extract from oocytes or HeLa cells and by the purified La-RNP complex. By purifying the S100 extract, these authors have obtained a fraction containing the 64 k polypeptide identified in the purified La-RNP but lacking the 50 k component. This purified S100 fraction was able to restore the inhibition of transcription as did the purified La-RNP and therefore it was conclude that the 50 k La-antigen is not a polymerase III transcription factor but that another protein present in the La-RNP is effectively required for transcription.

III. Anti-Jo-1 Antibody, $tRNA^{His}$ and Histidyl-tRNA Synthetase

The anti-Jo-1 antibody was first described by Nishikai and Reichlin (1980). It occurs in patients with polymyositis or dermatomyositis and is directed against an antigen which was found to be localized in the nucleus of mouse spleen cells by immunofluorescence staining. Attempts to biochemically characterize this antigen have led to the conclusion that it is a nuclear acidic protein of 150 k.

Using the same collection of sera, Rosa et al. (1983) have examined the pattern of RNA's present in an anti-Jo-1-immunoprecipitate obatined from a ^{32}P HeLa cell sonicate. In addition to some more variable species, they found a discrete species of RNA, also present in other cell extracts, which was identified as a $tRNA^{His}$ by RNA sequencing. In contrast to some other tRNA's (Epstein et al. 1982; Hardin et al. 1982), this Jo-1 RNA requires association with protein(s) to be immunoprecipitated since, only deproteinized by phenol extraction or proteinase K digestion, it is not recognized any more. However, all attempts to identify the protein(s) specifically associated with Jo-1 RNA have failed, and the Jo-1 RNA was found to coelute with the bulk of tRNA's through a Sephacryl S-200 column. At variance with the previous report from Nishikai and Reichlin (1980), the cellular localization of the Jo-1 antigen was found to be clearly cytoplasmic.

By identification of a possibly phosphorylated protein of 55 k associated with Jo-1 RNA, Mathews and Bernstein (1983) have postulated that the Jo-1 antigen might be the histidyl-tRNA synthetase and finally showed that the charging of tRNA with histidine is efficiently and specifically inhibited by anti-Jo-1 antibody added to a crude synthetase preparation. Moreover, when this synthetase preparation was depleted of the Jo-1 antigen by passing through immobilized anti-Jo-1 immunoglobulin, the charging of tRNA with histidine did not take place. Thus, the 55 k Jo-1 antigen might be a subunit of histidyl-tRNA synthetase which has been described as being a dimer of similar or identical subunits of about 64 k (Kane et al. 1978). More recently, Mathews and Bernstein (pers. comm.) have found that another aminoacyl-tRNA synthetase/tRNA complex reacts with an antibody different of anti-Jo-1 but also present in a myositis serum and Okada et al. (1984) have isolated from a patient with polymyositis a novel antibody precipitating ribonucleoprotein complex containing threonine tRNA.

D. Other Small RNP's of Characterized Function

I. RNAse P

RNAse P was first described as an enzymatic activity cleaving tRNA precursors to generate the 5' termini of mature tRNA's (for review, see Altman 1978). The most interesting feature of this enzyme is that it is an RNP, stable in cesium chloride (a feature shared by the core U snRNP's) whose RNA component is required for activity. Earlier reports argue that neither the 17.5 k protein nor the RNA have any activity by themselves but a functional enzyme can be reconstituted from inactive RNA and protein (Kole and Altman 1979, 1981; Kole et al. 1980). However, it appears from recent reports that the RNA moiety (M1 RNA) alone cleaves tRNA precursors in vitro and that the protein subunit acts as a cofactor that both enhances the rate of the cleavage and modifies the substrate specificity of the RNA-protein enzyme complex (Guerrier-Takada et al. 1983, 1984; Guerrier-Takada and Altman 1984). The M1 RNA component has now been cloned and sequenced (Reed et al. 1982). An RNAse P-like activity has also been described in eukaryotic systems (Bowman and Altman 1980; Akaboshi et al. 1980). Besides the peptidyl transferase activity of ribosomes, RNAse P represents the unprecedented situation of a very simple RNP endowed with a well-defined enzymatic activity involved with RNA processing.

II. The 5S RNP

The previtellogenic oocytes of *Xenopus laevis* contain large amounts of 5S RNA. About one half is associated with tRNA in a 42S particle containing two proteins of 40 k and 48 k, while the other half is bound to a 45 k protein forming an abundant 5S RNA-containing RNP migrating at 7S (Picard and Wegnez 1979). This protein is different from the one which is released with 5S RNA when ribosomes of mature oocytes are treated with EDTA, indicating that 5S RNA associates with two kinds of protein during the lifetime of the oocyte. Most interesting was the finding that the protein found in the storage particle described by Picard and Wegnez (1979) is actually the factor required for accurate transcription by RNA-polymerase III of 5S RNA genes whose promoter is internal to the 5S sequence (Engelke et al. 1980; Pelham and Brown 1980). This provides the basis for an elegant autoregulation model in which the transcription factor first promotes the production of 5S RNA until it is displaced by an excess of the latter to form the storage particle. As shown by immunohistochemistry, this storage particle, as well as the 42S particle, has an exclusively cytoplasmic localization (Mattaj et al. 1983).

III. The Signal Recognition Particle (SRP) and 7S RNA

The signal recognition particle (SRP) was first described as an 11S multi-subunits protein complex containing six polypeptides (72, 68, 54, 19, 14, and 9 k) which purified to homogeneity from a microsomal membranes salt wash (Walter and Blobel 1980). The SRP was able to restore the translocating activity of salt-washed microsomal membranes vesicles and was first named Signal Recognition Protein. When interacting with polysomes synthesizing secretory proteins, the SRP recognizes a signal peptide of 15 - 30 amino acids in length which is usually located at the NH_2 terminus of the pre-secretory protein and serves to arrest the translation. It then mediates the attachment of polysomes to microsomal membranes through a SRP receptor, releases the arrested translation and allows the secretory protein to be translocated across the lipid bilayer of the endoplasmic reticulum (Walter et al. 1981; Walter and Blobel 1981a,b; Meyer et al. 1982).

The high A_{260}/A_{280} ratio of their SRP preparation led Walter and Blobel (1982) to suspect the presence of a copurifying RNA component since it was sensitive to alkaline hydrolysis. RNAse treatment modifies the sedimentation behavior of the SRP and destroys its functional activity, allowing the conclusion that 7S-L RNA is a constitutive element of the SRP with respect to both structure and function. The "Signal Recognition Protein" thus became the "Signal Recognition Particle". Partial sequence analysis of the SRP RNA (Walter and Blobel 1982) unambiguously established its identity to the 7S-L RNA recently sequenced in rat

(Li et al. 1982) and man (Ullu et al. 1982b). As to the contribution of the 7S-L RNA to SRP function, two hypotheses have been considered (Walter and Blobel 1982). The first one views it as a matrix on which is assembled the SRP or its receptor. The second one suggests that 7S-L RNA could act through hydrogen-bonding interactions with either mRNA or rRNA. It seems that most cellular 7S-L RNA is engaged in SRP as it is nearly quantitatively precipitated by antibodies raised against SRP proteins and which neutralize its activity (Walter and Blobel, pers. comm.).

Two species of 7S RNA have now been characterized: 7S-L and 7S-K respectively corresponding to the scL and snK species described by Zieve and Penman (1976) and to the 7S and 7-3 RNA's of Reddy et al. (1981b). 7S-L RNA was first identified as a component of oncornavirus particles (Bishop et al. 1970) and soon detected in uninfected cells as well (Erikson et al. 1973; Larsen et al. 1973). It was shown to exist as two structural isomers (Larsen et al. 1974), one being mostly found in the viron, the other one predominating in the cell (Robert-Robin et al. 1974). 7S-L RNA sequences were obtained by sequencing either RNA from rat Novikoff hepatoma cells (Li et al. 1982) or cDNA clones from man (Ullu et al. 1982b) or mouse (Balmain et al. 1982). Hybridization experiments of the human clone to the RNA of rodents, birds, amphibians and echinoderms suggested a high degree of evolutionary conservation (Ullu et al. 1982).

Finally, 7S-L RNA was found to be the only small cellular RNA which is complementary to repeated Alu or B_1 family sequences (Weiner 1980; Ullu and Melli 1982; Ullu et al. 1982b; Balman et al. 1982). The 5' third of the 295 nucleotides long sequence is homologous to Alu and B_1 sequences (Li et al. 1982) whereas a central fragment is homologous to a new set of human middle-repetitive sequences (Ullu et al. 1982). This bipartite structure seems to be reflected in the SRP itself as demonstrated by nuclease digestion experiments (Gundelfinger et al. 1983). Homologies have also been noted between the 5' end 7S-L RNA and the La 4.5S RNA (Li et al. 1982).

IV. The 12S Termination Factor

In injection experiments of the histone gene repeat h22 DNA of sea urchin *Psammechinus miliaris* into the *Xenopus* oocyte nucleus, Stunnenberg and Birnstiel (1982) found that the yield of genuine H_3 mRNA 3' ends was dependent on the coinjection of a chromosomal salt wash fraction containing an active component with a sedimentation coefficient of about 12S. More recently, Galli et al. (1983) have demonstrated that the 12S nuclear termination component is a small nuclear RNP containing an RNA of about 60 nucleotides in length which is now referred as U_7 snRNA (Strub et al. 1984). Although this RNA, present in low abundance, has no sequence homologies to any other U snRNA's nor shows sequence complementarity to these RNA's as is the case between U_4 and U_6 snRNA's (Hashimoto and Steitz 1984; Bringmann et al. 1984), it can be extracted from cell homogenates by anti-Sm antibodies.

This suggests that the U_7 snRNP shares one or more proteins in common with other U snRNP's.

E. Concluding Remarks

Within the last few years, snRNA's and scRNA's have erupted from a status of oddities into one of elements crucial to many aspects of the cell life. During the first phase a considerable wealth of structural knowledge was accumulated which culminated by the establishment of primary as well as secondary structures for many of those species and especially the U snRNA's. Most instrumental in the onset of the second phase was the breakthrough discovery that U snRNP's are the antigens recognized by patients' sera suffering from rheumatic autoimmune diseases. This finding had two extraordinarily fertile consequences: first, it established their RNP status and conferred upon them important functional potentialities which are just starting to be unveiled; second, it aroused a considerable interest for the unexploited mine of autoimmune antibodies whose diversity has certainly not been fully appreciated yet. Although other RNP's, like the signal recognition particle (SRP), worked their way to stardom without the direct help of these antibodies, they all benefited from the intense research activity in the area of small RNA's.

As far as the U snRNP's are concerned, rather good evidence is currently accumulating in favor of the involvement of U_1 in the splicing of nuclear premessenger. No experimental clues exist concerning the role of U_2, $U_4 - U_6$, and U_5 snRNP's, although they share with U_1 snRNP a common core structure. It will be quite interesting to learn how such strikingly common structural features can concur to possibly different functions. Essentially different in nearly all respects, except for the presence of the typical U snRNA cap, is U_3 which also lacks an experimental handle. The recent availability of an autoimmune antibody specific for this individual snRNP will certainly be of great help to assign it a function.

With respect to the anti-La recognizable RNP's, it is quite interesting to note that the La antigen is able to bind a wide spectrum of nuclear as well as cytoplasmic RNA-polymerase III transcripts, some of which only transiently before handing them over to other antigens which might be even more important for their job in the cell.

The reader will have certainly realized how rapidly expanding is our knowledge in this field. By the time this review is published, much more data will have accumulated and some of the present ones will probably be outdated.

Acknowledgements. We would like to thank our collaborators which have contributed some of the data presented here. We are also very grateful to our many colleagues who have made their results available to us prior to publication. Thanks are also due to Ms. Fillot and Fize for typing the manuscript. The work from our laboratory was supported by grants from CNRS (ERA 482), INSERM, DGRST, Ligue Contre Le Cancer and Fondation pour la Recherche Médicale.

References

Agris, P.F., Kikuchi, Y., Gross, H.J., Takano, M., Sharp, G.C.: Characterization of the autoimmune antigenic determinant for ribonucleoprotein (RNP) antibody. Immunol. Commun. 13, 137-150 (1984)

Akaboshi, E., Guerrier-Takada, C., Altman, S.: Veal heart ribonuclease P has an essential RNA component. Biochem. Biophys. Res. Commun. 96, 831-837 (1980)

Akusjarvi, G., Mathews, M.B., Anderson, P., Vennstrom, B., Pettersson, U.: Structure of genes for virus-associated RNA_1 and RNA_2 of adenovirus type 2. Proc. Natl. Acad. Sci. USA 77, 2424-2428 (1980)

Alonso, A., Jorcano, J.L., Beck, E., Spiess, E.: Isolation and characterization of drosophila melanogaster U_2 small nuclear RNA genes. J. Mol. Biol. 169, 691-705 (1983)

Altman, S.: Transfer RNA biosynthesis. In: International review of Biochemistry (ed. B.C.F. Clark), vol. 17, pp. 19-44. Baltimore: University Park Press 1978

Alwine, J.C., Dhar, R., Khoury, G.: A small RNA induced late in simian virus 40 infection can associate with early viral mRNAs. Proc. Natl. Acad. Sci. USA 77, 1379-1383 (1980)

Assens, C., Graafland, H., Liautard, J.P., Sri-Widada, J., Brunel C., Jeanteur, P.: Relationship between snRNA species contained in nuclear antigens recognized by autoantibodies and the clinical profile in systemic rheumatism diseases. Biomedicine 36, 134-139 (1982a)

Assens, C., Liautard, J.P., Sri-Widada, J., Brunel, C., Jeanteur, P.: Highly purified snRNP's retain antigenic determinants towards anti Sm antibodies. Biochem. Biophys. Res. Commun. 106, 953-960 (1982b)

Bachellerie, J.P., Michot, B., Raynal, F.: Recognition signals for mouse pre-rRNA processing: A potential role for U_3 nucleolar RNA. Mol. Biol. Rep. 9, 79-86 (1983)

Balmain, A., Krumlauf, R., Vass, J.K., Birnie, G.D.: Cloning and characterization of the abundant cytoplasmic 7S RNA from mouse cells. Nucleic Acids Res. 10, 4259-4277 (1982)

Barque, J.P., Yeni, P., Peraudeau, L., Danon, F., Larsen, C.J.: Nuclear ribonucleoproteins recognized by human antinuclear antibodies in retrovirus-infected cells. Biochem. Biophys. Res. Commun. 99, 284-291 (1981)

Beck, E., Jorcano, J.L., Alonso, A.: Drosophila melanogaster U_1 and U_2 small nuclear RNA genes contain common flanking sequences. J. Mol. Biol. 173, 539-542 (1984)

Benecke, B.J., Penman, S.: A new class of small nuclear RNA molecules synthesized by a type 1 RNA polymerase in HeLa cells. Cell 12, 939-946 (1977)

Benecke, B.J., Penman, S.: Large species differences in the pattern of snPI RNA which can distinguish ape from human. J. Cell Biol. 80, 778-783 (1979)

Berezney, R., Coffey, D.S.: Identification of a nuclear protein matrix. Biochem. Biophys. Res. Commun. 60, 1410-1417 (1974)

Berget, S.M.: Are U_4 small nuclear ribonucleoproteins involved in polyadenylation? Nature (London) 309, 179-182 (1984)

Bernstein, L.B., Mount, S.M., Weiner, A.M.: Pseudogenes for human small nuclear RNA U_3 appear to arise by integration of self-primed reverse transcripts of the RNA into new chromosomal sites. Cell 32, 461-472 (1983)

Billings, P.B., Hoch, S.O.: Isolation of intact SM/RNP antigens from rabbit thymus. J. Immunol. 131, 347-351 (1983)

Billings, P.B., Hoch, S.O.: Characterization of U snRNA-associated proteins. J. Biol. Chem. 259, 12850-12856 (1984)

Billings, P.B., Allen, R.W., Jensen, F.C., Hoch, S.O.: Anti-RNP monoclonal antibodies derived from a mouse strain with Lupus-like autoimmunity. J. Immunol. 128, 1176-1180 (1982)

Bishop, M.M., Levinson, W.E., Quintrell, N., Sullivan, D., Fanshier, L., Jackson, J.: The low molecular weight RNAs of Rous sarcoma virus I. The 4S RNA. Virology 42, 182-195 (1970)

Blanchard, J.M.: hnRNP proteins interact with nascent adenovirus RNA. Biochem. Biophys. Res. Commun. 97, 524-529 (1980)

Bowman, E.J., Altman, S.: Identification of ribuonuclease P activity from chicken embryos. Biochim. Biophys. Acta 613, 439-447 (1980)

Bozzoni, I., Amaldi, F., Annesi, F., Beccari, E., Fragapane, P., Pierandrei-Amaldi, P.: Splicing of xenopus laevis ribosomal protein RNAs is in vivo inhibited by antisera to snRNP. J. Mol. Biol. 180, 1173-1178 (1984)

Branlant, C., Krol, A., Ebel, J.P., Lazar, E., Gallinaro, H., Jacob, M., Sri-Widada, J., Jeanteur, P.: Nucleotide sequences of nuclear U_1A RNAs from chicken, rat and human. Nucleic Acids Res. 8, 4143-4154 (1980)

Branlant, C., Krol, A., Ebel, J.P., Gallinaro, H., Lazar, E., Jacob, M.: The conformation of chicken, rat and human U_1A RNAs in solution. Nucleic Acids Res. 9, 841-852 (1981)

Branlant, C., Krol, A., Ebel, J.P., Lazar, E., Haendler, B., Jacob, M.: U_2 RNA shares a structural domain with U_1, U_4 and U_5 RNAs. EMBO J. 1, 1259-1265 (1982)

Branlant, C., Krol, A., Lazar, E., Haendler, B., Jacob, M., Galego-Dias, L., Pousada, C.: High evolutionary conservation of the secondary structure of certain nucleotide sequences of U_5 RNA. Nucleic Acids Res. 11, 8359-8367 (1983)

Breathnach, B., Chambon, P.: Organization and expression of eucaryotic split genes coding for proteins. Annu. Rev. Biochem. 50, 349-383 (1981)

Bringmann, P., Reuter, R., Rinke, J., Appel, B., Bald, R., Luhrmann, R.: 5' terminal caps of snRNAs are accessible for reaction with 2,2,7-trimethyl guanosine-specific antibody in intact snRNPs. J. Biol. Chem. 258, 2745-2747 (1983a)

Bringmann, P., Rinke, J., Appel, B., Reuter, R., Luhrmann, R.: Purification of snRNPs U_1, U_2, U_4, U_5 and U_6 with 2,2,6-trimethylguanosine-specific antibody and defition of their constituent proteins reacting with anti-Sm and anti-(U_1)RNP antisera. EMBO J. 2, 1129-1135 (1983b)

Bringmann, P., Appel, B., Rinke, J., Reuter, R., Theissen, H., Luhrmann, R.: Evidence for the existence of snRNAs U_4 and U_6 in a single ribonucleoprotein complex and for their association by intermolecular base pairing. EMBO J. 3, 1357-1363 (1984)

Brown, A., Marzluff, W.: Biosynthesis of low molecular weight RNA in mouse myeloma cells. Biochim. Biophys. Acta 521, 662-676 (1978)

Brunel, C., Sri-Widada, J., Lelay, M.N., Jeanteur, P., Liautard, J.P.: Purification and characterization of a simple ribonucleoprotein particle containing small nucleoplasmic RNAs (snRNP) as a subject of RNP containing heterogeneous nuclear RNA (hnRNP) from HeLa cells. Nucleic Acids Res. 9, 815-830 (1981)

Busch, H., Reddy, R., Rothblum, L., Choi, Y.E.: SnRNA, SnRNPs and RNA processing. Annu. Rev. Biochem. 51, 617-654 (1982)

Calvet, J.P., Pederson, T.: Base-pairing interactions between small nuclear RNAs and nuclear RNA precursors as revealed by psoralen cross-linking in vivo. Cell 26, 363-370 (1981)

Calvet, J.P., Meyer, L.M., Pederson, T.: Small nuclear RNA U2 is base-paired to heterogeneous nuclear RNA. Science 217, 465-458 (1982)

Card, C.O., Morris, G.F., Brown, D.T., Marzluff, W.F.: Sea urchin small nuclear RNA genes are organized in distinct tandemly repeating units. Nucleic Acids Res. 10, 7677-7688 (1982)

Cervera, M., Dreyfuss, G., Penman, S.: Messenger RNA is translated when associated with the cyto-skeletal framework in normal and VSV-infected HeLa cells. Cell 23, 113-120 (1981)

Chandrasekharappa, S.C., Smith, J.H., Eliceiri, G.: Biosynthesis of small nuclear RNAs in human cells. J. Cell Physiol. 117, 169-174 (1983)
Crouch, R.J., Kanaya, S., Earl, P.L.: A model for the involvement of the small nucleolar RNA (U_3) in processing eukaryotic ribosomal RNA. Molec. Biol. Rep. 9, 75-78 (1983)
Dambaugh, T., Beisel, C., Hummel, M., King, W., Fennewald, S., Cheung, A., Heller, M., Raab-Traub, N., Krieff, E.: Epstein-barr virus (B95-8) DNA. VII: molecular cloning and detailed mapping. Proc. Natl. Acad. Sci. (USA) 77, 2999-3003 (1980)
Daskal, Y.: Perichromatin granules. In: The Cell Nucleus (ed. H. Busch), Vol. VIII, pt. A, pp. 117-137. London-New York: Academic Press 1981
Daskal, Y., Komaromy, L., Busch, H: Isolation and partial characterization of perichromatin granules. A unique class of nuclear RNP particles. Exp. Cell Res. 126, 39-46 (1980)
Deimel, B., Louis, Ch., Sekeris, C.E.: The presence of small molecular weight RNAs in nuclear ribonucleoprotein particles carrying hnRNAs. FEBS Lett. 73, 80-84 (1977)
Denison, R.A., Weiner, A.M.: Human U_1 RNA pseudogenes may be generated by both DNA and RNA mediated mechanisms. Mol. Cell. Biol. 2, 815-828 (1982)
Denison, R.A., Vanarsdell, S.W., Bernstein, L.B., Weiner, A.M.: Abundant pseudogenes for small nuclear RNAs are dispersed in the human genome. Proc. Natl. Acad. Sci. (USA) 78, 810-814 (1981)
De Robertis, E.M., Lienhard, S., Parisot, R.F.: Intracellular transport of microinjected 5S and small nuclear RNAs. Nature (London) 295, 572-577 (1982)
Dingman, C.W., Peacock, A.C.: Analytical studies on nuclear ribonucleic acid using polyacrylamide gel electrophoresis. Biochemistry 7, 659-668 (1968)
Douvas, A.S.: Auto antibodies occuring in two different rheumatic diseases react with the same nuclear ribonucleoprotein particle. Proc. Nátl. Acad. Sci. USA 79, 5401-5405 (1982)
Douvas, A.S., Stumph, W.E., Reyes, P., Tan, E.M.: Isolation and characterization of nuclear ribonucleoprotein complexes using human anti-nuclear ribonucleoprotein antibodies. J. Biol. Chem. 254, 3608-3616 (1979)
Eliceiri, G.L.: Short-lived, small RNAs in the cytoplasm of HeLa cells. Cell 3, 11-14 (1974)
Eliceiri, G.L.: Sensitivity of low molecular weight RNA snythesis to UV radiation. Nature (London) 279, 80-81 (1979)
Eliceiri, G.L.: Formation of low molecular weight RNA species in HeLa cells. J. Cell. Physiol. 102, 199-207 (1980)
Eliceiri, G.L., Gurney, Jr. T.: Subcellular location of precursors to small nuclear RNA species C and D and of newly synthesized 5S RNA in HeLa cells. Biochem. Biophys. Res. Commun. 81, 915-919 (1978)
Eliceiri, G.L., Sayavedra, M.S.: Small RNAs in the nucleus and cytoplasm of HeLa cells. Biochem. Biophys. Res. Commun. 72, 507-512 (1976)
Emerson, S.V., Dierks, P.M., Parsons, J.T.: In vitro synthesis of a unique RNA species by a T particle of vesicular stomatitis virus. J. Virol. 23, 708-716 (1977)
Engelke, D.R., Ng, S.Y., Shastry, B.S., Roeder, R.G.: Specific interaction of a purified transcription factor with an internal control region of 5S RNA genes. Cell 19, 717-728 (1980)
Enger, M.D., Walters, R.A.: Isolation of low molecular weight, methylated ribonucleic acids from 10S to 30S particles of chinese hamster cell fractions. Biochemistry 9, 3551-3562 (1970)
Epstein, P., Reddy, R., Henning, D., Busch, H.: The nucleotide sequence of nuclear U6 (4.7S) RNA. J. Biol. Chem. 255, 8901-8906 (1980)
Epstein, P., Reddy, R., Busch, H.: Site-specific cleavage by T1 RNAse of U-1 RNA in U-1 ribonucleoprotein particles. Proc. Natl. Acad. Sci. USA 78, 1562-1566 (1981)

Epstein, P., Lidsky, M., Reddy, R., Tan, E., Busch, H.: Identification of three different anti-4S RNA sera associated with autoimmune disease. Biochem. Biophys. Res. Commun. 109, 548-555 (1982)

Erikson, E., Erikson, R.L., Henry, B., Pace, M.R.: Comparison of oligonucleotides produced by RNAse T_1 digestions of 7S RNA from avian and murine oncornavirus and from uninfected cells. Virology 53, 40-46 (1973)

Fakan, S., Leser, G., Martin, T.E.: Ultractructural distribution of nuclear ribonucleoproteins as visualized by immunocytochemistry on thin sections. J. Cell. Biol. 98, 358-363 (1984)

Fisher, D.E., Conner, G.E., Reeves, W.H., Blobel, G., Kunkel, H.G.: Synthesis and assembly of human small nuclear ribonucleoproteins generated by cell-free translation. Proc. Natl. Acad. Sci. USA 80, 6356-6360 (1983)

Fisher, D.E., Reeves, W.H., Conner, G.E., Blobel, G., Kunkel, H.G.: Pulse labelling of small nuclear ribonucleoproteins in vivo reveals distinct patterns of antigen recognition by human autoimmunes antibodies. Proc. Natl. Acad. Sci. USA 81, 3185-3189 (1984)

Flytzanis, C., Alonso, A., Louis, Ch, Krieg, L., Sekeris, C.E.: Association of small nuclear RNA with hnRNA isolated from nuclear RNP complexes carrying hnRNA. FEBS Lett. 96, 201-206 (1978)

Fowlkes, D.M., Shenk, T.: Transcriptional control regions of the adenovirus VAI RNA gene. Cell 22, 405-413 (1980)

Fradin, A., Jove, R., Hemenway, C., Keiser, H.D., Manley, J.L., Prives, C.: Splicing pathways of SV40 mRNAs in X. Laevis oocytes differ in their requirements for snRNPs. Cell 37, 927-936 (1984)

Francoeur, A.M., Mathews, M.B.: Interaction between VA RNA and the lupus antige La: formation of a ribonucleoprotein particle in vitro. Proc. Natl. Acad. Sci. USA 79, 6772-6776 (1982)

Frederiksen, S., Hellung-Larsen, P.: Precursors to small molecular weight RNA components. FEBS Lett. 58, 374-378 (1975)

Frederiksen, S., Hellung-Larsen, P., Gram Jensen, E.: The differential inhibitory effect of α-amanitin on the synthesis of low molecular weight RNA components in BHK cells. FEBS Lett. 87, 227-231 (1978)

Frederiksen, S., Flodgard, H., Hellung-Larsen, P.: Subcellular localization of low-molecular weight RNA components in rat liver. Biochem. J. 193, 743-748 (1981)

Galli, G., Hofstetter, H., Birnstiel, M.L.: Two conserved sequence blocks within eucaryotic tRNA genes are major promoter elements. Nature (London) 294, 626-631 (1981)

Galli, G., Hofstetter, H., Stunnenberg, H.G., Birnstiel, M.: Biochemical complementation with RNA in the xenopus oocyte: a small RNA is required for the generation of 3' histone mRNA termini. Cell 34, 823-828 (1983)

Gallinaro, H., Jacob, M.: An evaluation of small nuclear RNA in hnRNP. FEBS Lett. 104, 176-182 (1979)

Gallinaro, H., Jacob, M.: The status of small nuclear RNA in the ribonucleoprotein fibrils containing heterogeneous nuclear RNA. Biochim. Biophys. Acta 652, 109-120 (1981)

Gallinaro, H., Gattoni, R., Stevenin, J., Jacob, M.: Small nuclear RNA and VA RNA in nuclear ribonucleoprotein fibrils from adenovirus-2 infected HeLa cells. Biochem. Biophys. Res. Commun. 95, 20-26 (1980)

Gallinaro, H., Lazar, E., Jacob, M., Krol, A., Branlant, C.: Small RNAs in hnRNP fibrils and their possible function in splicing. Molec. Biol. Rep. 7, 31-39 (1981)

Gallinaro-Matringe, H., Stevenin, J., Jacob, M.: Salt dissociation of nuclear particles containing DNA-like RNA. Distribution of phosphorylated and non-phosphorylated species. Biochemistry 14, 2547 2554 (1975)

Gattoni, R., Stevenin, J., Devilliers, G., Jacob, M.: Size heterogeneity of monoparticles from nuclear ribonucleoproteins containing premessenger RNA. FEBS Lett. 90, 318-323 (1978)

Gojobori, T., Nei, M.: Inter-RNA homology and possible roles of small RNAs. J. Mol. Evol. 17, 245-250 (1981)

Gottesfeld, J.M., Andrews, D.L., Hoch, S.O.: Association of an RNA polymerase III transcription factor with a ribonucleoprotein complex recognized by autoimmune sera. Nucleic Acids Res. 12, 3185-3200 (1984)

Gram Jensen, E., Hellung-Larsen, P., Frederiksen, S.: Synthesis of low molecular weight weight RNA components A, C and D by polymerase II in α-amanitin-resistant hamster cells. Nucleic Acids Res. 6, 321-330 (1979)

Guerrier-Takada, C., Altman, S.: Catalytic activity of an RNA molecule prepared by transcription in vitro. Science 223, 285-286 (1984)

Guerrier-Takada, C., Gardiner, K., Marsh, T., Pace, N., Altman, S.: The RNA moiety of ribonuclease P is the catalytic subunit of the enzyme. Cell 35, 849-847 (1983)

Guerrier-Takada, C., McClain, W.H., Altman, S.: Cleavage of tRNA precursors by the RNA subunit of E. coli ribonuclease P (M1 RNA) is influenced by 3'-proximal CCA in the substrates. Cell 38, 219-224 (1984)

Guilfoyle, R., Weinmann, R.: Control region for adenovirus VA RAN transcription. Proc. Natl. Acad. Sci. USA 78, 3378-3382 (1981)

Guimont-Ducamp, Ch., Sri-Widada, J., Jeanteur, Ph.: Occurence of small molecular weight RNAs in HeLa nuclear ribonucleoprotein particles containing hnRNA (hnRNP). Biochimie 59, 755-758 (1977)

Gundelfinger, E.D., Krause, E., Melli, M., Dobberstein, B.: The organization of the 7SL RNA in the signal recognition particle. Nucleic Acids Res. 11, 7363-7374 (1983)

Gurney, T. Jr., Eliceiri, G.L.: Intracellular distribution of low molecular weight RNA species in HeLa cells. J. Cell Biol. 87, 398-403 (1980)

Habets, W.J., De Rooij, D.J., Salden, M.H., Verhagen, A.P., Van Eekelen, C.A., Van de Putte, L.B., Van Venrooij, W.J.: Antibodies against distinct nuclear matrix proteins are characteristic for mixed connective tissue disease. Clin. Exp. Immunol. 54, 265-276 (1983a)

Habets, W.J., Den Brok, J.H., Boerbooms, A.M.Th., Van de Putte, L.B.A., Van Venrooij, W.J.: Characterisation of the SS-B (La) antigen in adenovirus-infected and uninfected HeLa cells. EMBO J. 2, 1625-1631 (1983b)

Hammarstrom, K., Westin, G., Pettersson, U.: A pseudogene for human U4 RNA with a remarkable structure. EMBO J. 1, 737-739 (1982)

Harada, F., Ikawa, Y.: A new series of RNAs associated with the genome of spleen focus forming virus (SFFV) and poly(A)-containing RNA from SFFV-infected cells. Nucleic Acids Res. 7, 895-908 (1979)

Harada, F., Kato, N.: Nucleotide sequences of 4.5S RNAs associated with poly(A)-containing RNAs of mouse and hamster cells. Nucleic Acids Res. 8, 1273-1285 (1980)

Harada, F., Kato, N., Hoshino, H.: Series of 4.5S RNAs associated with poly(A)-containing RNAs of rodent cells. Nucleic Acids Res. 7, 909-917 (1979)

Harada, F., Kato, N., Nishimura, S.: The nucleotide sequence of nuclear 4.8S RNA of mouse cells. Biochem. Biophys. Res. Commun. 95, 1332-1340 (1980)

Hardin, J.A., Rahn, D.R., Shen, C., Lerner, M.R., Wolin, S.L., Rosa, M.D., Steitz, J.A.: Antibodies from patients with connective tissue diseases bind specific subsets of cellular RNA-protein particles. J. Clin. Invest. 70, 141-147 (1982)

Harris, B., Roeder, R.G.: Structural relation-ships of low molecular weight viral RNAs synthesized by RNA polymerase III in nuclei from adenovirus-2 infected cells. J. Biol. Chem. 253, 4120-4127 (1978)

Hashimoto, C., Steitz, J.A.: Sequential association of nucleolar 7-2RNA with two different auto antigens. J. Biol. Chem. 258, 1379-1382 (1983)

Hashimoto, C., Steitz, J.A.: U4 and U_6 RNAs coexist in a single small nuclear ribonucleoprotein particle. Nucleic Acids Res. 12, 3283-3293 (1984)

Hayashi, K.: Organization of sequences related to U_6 RNA in the human genome. Nucleic Acids Res. 9, 3379-3388 (1981)

Haynes, S.R., Jelinek, W.R.: Low molecular weight RNAs transcribed in vitro by RNA polymerase III from Alu-type dispersed repeats in chinese hamster DNA are also found in vivo. Proc. Natl. Acad. Sci. USA 78, 6130-6134 (1981)

Hellung-Larsen, P.: Low molecular weight RNA components in eukaryotic cells. Cophenhagen-Arhus-Odense (FADL's Forlag 1977)

Hellung-Larsen, P., Kulamowicz, I., Frederiksen, S.: Synthesis of low molecular weight RNA components in cells with a temperature sensitive polymerase II. Biochim. Biophys. Acta 609, 201-204 (1980)

Hellung-Larsen, P., Gram Jensen, E., Frederiksen, S.: Effect of 5,6-dichloro-1-β-D-ribofuranosylbenzimidazole on the synthesis of low molecular weight RNA components. Biochem. Biophys. Res. Commun. 99, 1303-1310 (1981)

Hendrick, J.P., Wolin, S.L., Rinke, J., Lerner, M.R., Steitz, J.A.: Ro small cytoplasmic ribonucleoproteins are a subclass of La ribonucleoproteins: further characterization of the Ro and La small ribonucleoproteins from uninfected mammalian cells. Mol. Cel. Biol. 1, 1138-1149 (1981)

Hernandez, N., Keller, W.: Splicing of in vitro synthesized messenger RNA precursors in HeLa cell extracts. Cell 35, 89-99 (1983)

Hinterberger, M., Pettersson, I., Steitz, J.A.: Isolation of small ribonucleoproteins containing U_1, U_2, U_4, U_5 and U_6 RNAs. J. Biol. Chem. 258, 2604-2613 (1983)

Hoch, S.O., Billings, P.B.: Characterization of the La/SS-B antigen from several Mammalian sources. J. Immunol. 133, 1397-1403 (1984)

Honda, B.M., Roeder, R.G.: Association of a 5S gene transcription factor with 5S RNA and altered levels of the factor during cell differentiation. Cell 22, 119-126 (1980)

Howard, E.F.: Small nuclear RNA molecules in nuclear ribonucleoprotein complexes from mouse erythroleukemia cells. Biochemistry 17, 3228-3232 (1978)

Hutchinson, M.A., Hunter, T., Eckhart, W.: Characterization of T antigens in polyoma-infected and transformed cells. Cell 15, 65-77 (1979)

Jelinek, W., Leinwand, L.: Low molecular weight RNAs hydrogen bonded to nuclear and cytoplasmic poly(A)-terminated RNA from cultured chinese hamster ovary cells. Cell 15, 205-214 (1978)

Jelinek, W.R., Toomey, T.P., Leinwand, L., Duncan, C.H., Biro, P.A., Choudary, P.V., Weissman, S.M., Rubin, C.M., Houck, C.M., Deininger, P.L., Schmid, C.W.: Ubiquitous, interspersed repeated sequences in mammalian genomes. Proc. Natl. Acad. Sci. USA 77, 1398-1402 (1980)

Kane, S.M., Vurgrincic, C., Finbloom, D.S., Smith, D.W.E.: Purification and some properties of the histidyl tRNA synthetase from the cytosol of rabbit reticulocytes. Biochemistry 17, 1509-1514 (1978)

Kato, N., Harada, F.: Nucleotide sequence of nuclear 5S RNA of mouse cells. Biochem. Biophys. Res. Commun. 99, 1468-1476 (1981a)

Kato, N., Harada, F.: Nucleotide sequence of nuclear 5.7S RNA of mouse cells. Biochem. Biophys. Res. Commun. 99, 1477-1485 (1981b)

Kato, N., Hoshino, H.O., Harada, F.: Nucleotide sequence of 4.5S RNA (C8 or hY5) from HeLa cells. Biochem. Biophys. Res. Commun. 108, 363-370 (1982)

Kennedy, D.S., Siegel, E., Heywood, S.M.: Purification of myosin mRNP translational control RNA and its inhibition of myosin and globin messenger translation. FEBS Lett. 90, 209-214 (1978)

Kinlaw, C.S., Dusing-Swartz, S.K., Berget, S.M.: Human U_1 and U_2 small nuclear ribonucleoproteins contains common and unique polypeptides. Mol. Cell. Biol. 2, 1159-1166 (1982)

Kinlaw, C.S., Robberson, B.L., Berget, S.M.: Fractionation and characterization of human small nuclear ribonucleoproteins containing U_1 and U_2 RNAs. J. Biol. Chem. 258, 7181-7189 (1983)

Kole, R., Altman, S.: Reconstitution of RNAse P activity from inactive RNA and protein. Proc. Natl. Acad. Sci. USA 76, 3795-3799 (1979)

Kole, R., Altman, S.: Properties of purified ribonuclease P from Escherichia Coli. Biochemistry 20, 1902-1906 (1981)

Kole, R., Baer, M.F., Stark, B.C., Altman, S.: E. Coli RNAse P has a required RNA component in vivo. Cell 19, 881-887 (1980)

Krämer, A., Keller W., Appel, B., Kührman, R.: The 5' terminus of the RNA moiety of U_1 small nuclear ribonucleoprotein particles is required for the splicing of messenger RNA precursors. Cell 38, 299-307 (1984)

Krol, A., Branlant, C., Lazar, E., Gallinaro, H., Jacob, M.: Primary and secondary structures of chicken, rat and man nuclear U_4 RNAs. Homologies with U_1 and U_5 RNAs. Nucleic Acids Res. 9, 2699-2716 (1981a)

Krol A., Gallinaro, H., Lazar, E., Jacob, M., Branlant, C.: The nuclear 5S RNA from chicken, rat and man. U_5 RNAs are incoded by multiple genes. Nucleic Acids Res. 9, 769-787 (1981b)

Krol, A., Ebel, J.P., Rinke, J., Luhrman, R.: U_1, U_2 and U_5 small nuclear RNAs are found in plant cells. Complete nucleotide sequence of the U_5 RNA family from pea nuclei. Nucleic Acids Res. 11, 8583-8594 (1983)

Kurilla, M.G., Keene, J.D.: The leader RNA of vesicular stomatitis virus is bound by a cellular protein reactive with anti-La lupus antibodies. Cell 34, 837-845 (1983)

Larsen, C.J., Galibert, F., Lelong, J.C., Boiron, M.: Characterisation dans la fraction nucléaire de cellules KB d'un ARN métaboliquement stable de faible poids moléculaire. C.R. Acad. Sci. 264, 1523-1526 (1967)

Larsen, C.J., Galibert, F., Hampe, A., Boiron, M.: Etude des RNA nucléaires de faible poids moléculaire de la cellule KB. Bull. Soc. Chim. Biol. 51, 649-668 (1969)

Larsen, C.J., Emanoil-Rabicovitch, R., Samso, A., Robin, J., Tavitian, A., Boiron, M.: Presence of two 8S RNA components in mouse sarcoma virus (Moloney). Virology 54, 552-556 (1973)

Larsen, C.J., Ravicovitch, R.E., Hampe, A., Mauchauffe, M., Bazilier, M., Robert-Robin, J.: Identity of the two 8S RNA components of the mouse sarcoma virus (Molonay). Nucleic Acids Res. 1, 849-854 (1974)

Lazar, E., Jacob, M., Krol, A., Branlant, C.: Accessibility of U_1 RNA to base pairing with a single stranded DNA fragment mimicking the intron extremities at the splice junction. Nucleic Acids Res. 10, 1193-1201 (1982)

Lerner, E.A., Lerner, M.R., Janeway, C.A., Steitz, J.A.: Monoclonal antibodies to nucleic acid-containing cellular constituents: probes for molecular biology and autoimmune disease. Proc. Natl. Acad. Sci. USA 78, 2737-2741 (1981)

Lerner, M.R., Steitz, J.A.: Antibodies to small nuclear RNAs complexed with proteins are produced by patients with systemic lupus erythematosus. Proc. Natl. Acad. Sci. USA 76, 5495-5499 (1979)

Lerner, M.R., Steitz, J.A.: Snurps and Scyrps. Cell 25, 298-300 (1981)

Lerner, M.R., Boyle, J., Mount, S., Wolin, S., Steitz, J.A.: "Are snRNPs involved in splicing?". Nature (London) 283, 220-224 (1980)

Li, W.Y., Reddy, R., Henning, D., Epstein, P., Busch, H.: Nucleotide sequence of 7S RNA. J. Biol. Chem. 257, 5136-5142 (1982)

Liautard, J.P., Sri-Widada, J., Brunel, C., Jeanteur, P.: Structural organization of ribonucleoproteins containing small nuclear RNAs from HeLa cells. J. Mol. Biol. 162, 623-643 (1982)

Lin, W.L., Pederson, T.: Ribonucleoprotein organisation of eukaryotic RNA. Structure of U_1 small nuclear RNP. J. Mol. Biol. 180, 947-960 (1984)

Liu, M.H., Reddy, R., Henning, D., Spector, D., Bush, H.: Primary and secondary structure of dinoflagellate U_5 small nuclear RNA. Nucleic Acids Res. 12, 1529-1542 (1984)

Long, B.H., Ochs, R.L.: Nuclear matrix hnRNA and snRNA in friend erythroleukemia nuclei depleted of chromatin by low ionic strength EDTA. Biol. Cell 48, 89-98 (1983)

Long, B.H., Schrier, W.H.: Isolation from friend erythroleukemia cells of an RNAse-sensitive nuclear matrix fibril fraction containing hnRNA and snRNA. Biol. Cell 48, 99-108 (1983)

Lund, E., Dahlberg, J.E.: True genes for human U_1 small nuclear RNA. Copy number, polymorphism, and methylation. J. Biol. Chem. 259, 2013-2021 (1984)

Lund, E., Bostock, C., Robertson, M., Christie, S., Mitchen, J.L., Dahlberg, J.E.: U_1 small nuclear RNA genes are located on human chromosoma 1 and are expressed in mouse. Human hybrid cells. Mol. Cell. Biol. 3, 2211-2220 (1983)

Madore, S.J., Wieben, E.D., Pederson, T.: Intracellular site of U_1 small nuclear RNA processing and ribonucleoprotein assembly. J. Cell Biol. 98, 188-192 (1984a)

Madore, S.J., Wieben, E.D., Pederson, T.: Eukaryotic small ribonucleoproteins. Anti-La human autoantibodies react with U_1 RNA protein complexes. J. Biol. Chem. 259, 1929-1933 (1984b)

Madore, S.J., Wieben, E.D., Kunkel, G.R., Pederson, T.: Precursors of U_4 small nuclear RNA. J. Cell Biol. 99, 1140-1144 (1984c)

Manley, J.L., Fire, A., Cano, A., Sharp, P.A., Gefter, M.L.: DNA dependent transcription of adenovirus genes in a soluble whole-cell extract. Proc. Natl. Acad. Sci. USA 77, 3855-3859 (1980)

Manser, T., Gesteland, R.: Human U_1 loci: genes for human U_1 RNA have dramatically similar genomic environments. Cell 29, 257-264 (1982)

Manuelidis, L.: Nucleotide sequence defintion of a major human repeated DNA, the Hind III 1.9 Kb family. Nucleic Acids Res. 10, 3211-3219 (1982)

Marzluff, W.F., White, E.L., Benjamin, R., Huang, R.C.C.: Low molecular weight RNA species from chromatin. Biochemistry 14, 3715-3724 (1975)

Marzluff, W.F., Brown, D.T., Lobo, S., Wang, S.S.: Isolation and characterization of two linked mouse U_{1b} small nuclear RNA genes. Nucleic Acids Res. 11, 6255-6270 (1983)

Mathews, M.B., Bernstein, R.M.: Myositis autoantibody inhibits histidine-tRNA synthetase: A model for autoimmunity. Nature (London) 104, 177-179 (1983)

Mathews, M.B., Pettersson, U.: The low molecular weight RNAs of adenovirus 2 infected cells. J. Mol. Biol. 119, 293-328 (1978)

Mattaj, I.W., Zeller, R.: Xenopus laevis U_2 snRNA genes: tandemly repeated transcription units sharing 5' and 3' flanking homology with other RNA polymerase II transcribed genes. EMBO J. 2, 1883-1891 (1983)

Mattaj, I.W., Lienhard, S., Zeller, R., De Robertis, E.M.: Nuclear exclusion of transcription factor IIIA and the 42S particle transfer RNA-binding protein in xenopus oocytes: a possible mechanism for gene control? J. Cell Biol. 97, 1261-1265 (1983)

Mattaj, I.W., De Robertis, E.M.: Nuclear segregation of U_2 snRNA requires binding of specific snRNP proteins. Cell 40, 111-118 (1985)

Matter, L., Schopfer, K., Wilhelm, J.A., Nyffenegger T., Parisot, R.F., De Robertis, E.M.: Molecular characterization of ribonucleoprotein antigens bound by antinuclear antibodies. Arthritis Rheum. 25, 1278-1283 (1982)

Maxwell, E.S., Maundrell, K., Scherrer, K.: Two low molecular weight nuclear RNAs, isolated from avian erythroblast nuclear ribonucleoprotein complexes, hybridize to duck pre-messenger and globin messenger RNA. Biochem. Biophys. Res. Commun. 97, 875-882 (1980)

Maxwell, E.S., Maundrell, K., Puvion-Dutilleul, F., Scherrer, K.: Distribution of low-molecular weight RNAs in avian erythroblast nuclear ribonucleoprotein complexes associated with pre-messenger RNA. Eur. J. Biochem. 113, 233-247 (1981)

Meyer, D.I., Krause, E., Dobberstein, B.: Secretory protein translocation across membranes-the role of the "docking protein". Nature (London) 297, 647-650 (1982)

Miller, T.E., Huang, C.Y., Pogo, A.O.: rat liver nuclear skeleton and small molecular weight RNA species. J. Cell Biol. 76, 692-704 (1978)

Mimori, T., Hinterberger, M., Pettersson, I., Steitz, J.A.: Autoantibodies to the U_2 small nuclear ribonucleoprotein in a patient with scleroderma-polymyositis overlap syndrome. J. Biol. Chem. 259, 560-565 (1984)

Monneron, A., Bernhard, W.: Fine structural organization of the interphase nucleus in some mammalian cells. J. Ultrastruc. Res. 27, 266-288 (1969)

Monstein, H.J., Philipson, L.: The conformation of adenovirus VAI-RNA in solution. Nucleic Acids Res. 9, 4239-4250 (1981)

Monstein, H.J., Westin, G., Philipson, L., Pettersson, U.: A candidate gene for human U_1 RNA. EMBO J. 1, 133-137 (1982)

Moore, C.L., Sharp, P.A.: Site-specific polyadenylation in a cell-free reaction. Cell 36, 581-591 (1984)

Mount, S.M.: A catalogue of splice sequences. Nucleic Acids Res. 10, 459-473 (1982)

Mount, S.M., Steitz, J.A.: Sequence of U_1 RNA from Drosophila melanogaster: implications for U_1 secondary structure and possible involvement in splicing. Nucleic Acids Res. 9, 6351-6368 (1981)

Mount, S.M., Pettersson, I., Hinterberger, M., Karmas, A., Steitz, J.A.: The U_1 small nuclear RNA-protein complex selectively binds a 5' splice site in vitro. Cell 33, 509-518 (1983)

Muramatsu, M., Hodnett, J.L., Busch, H.: Base composition of fractions of nuclear and nucleolar ribonucleic acid obtained by sedimentation and chromatography. J. Biol. Chem. 241, 1544-1550 (1966)

Murphy, J.T., Burgess, R.R., Dahlberg, J.E., Lund, E.: Transcription of a gene for human U_1 small nuclear RNA. Cell 29, 265-274 (1982)

Murray, V., Holliday, R.: Mechanism for RNA splicing of gene transcripts. FEBS Lett. 106, 5-7 (1979)

Myslinski, E., Branlant, C., Wieben, E.D., Pederson, T.: The small nuclear RNA's of Drosophila. J. Mol. Biol. 180, 927-945 (1984)

Nakao, Y., Mukai, R., Kabashima, T., Hamaguchi, H., Kashiwagi, H., Okada, N.: A novel antibody which precipitates 7.5S RNA is isolated from a patient with autoimmune disease. Biochem. Biophys. Res. Commun. 109, 1332-1338 (1982)

Nishikai, M., Reichlin, M.: Heterogeneity of precipitating antibodies in polymyositis and dermatomyositis. Characterization of the Jo-1 antibody system. Arthritis Rheum. 23, 881-888 (1980)

Northemann, W., Scheurlen, M., Gross, V., Heinrich, P.C.: Circular dichroism of ribonucleoprotein complexes from rat liver nuclei. Biochem. Biophys. Res. Commun. 76, 1130-1137 (1977)

Northway, J.D., Tan, E.M.: Differentiation of antinuclear antibodies giving speckled staining patterns in immunofluorescence. Clin. Immunol. Immunopathol. 1, 140-154 (1972)

O'Farrell, P.Z., Goodman, H., O'Farrell, P.H.: High resolution two dimensional electrophoresis of basic as well as acidic proteins. Cell 12, 1133-1142 (1977)

Ohshima, Y., Itoh, M., Okada, N., Miyata, T.: Novel models for RNA splicing that involve a small nuclear RNA. Proc. Natl. Acad. Sci. USA 78, 4471-4474 (1981a)

Ohshima, Y., Okada, N., Tani, T., Itoh, Y., Itoh, M.: Nucleotide sequences of mouse genomic loci including a gene or pseudogene for U_6 (4.8S) nuclear RNA. Nucleic Acids Res. 9, 5145-5158 (1981b)

Okada, N., Mukai, R., Harada, F., Kabashima, T., Nakao, Y., Yamane, K., Ohshima, Y., Sakamoto, K., Itoh, M., Kashiwagi, H., Hamaguchi, H.: Isolation of a novel antibody, which precipitates ribonucleoprotein complex containing threonine tRNA from a patient with polymyositis. Eur. J. Biochem. 139, 425-429 (1984)

Padgett, R.A., Mount, S.M., Steitz, J.A., Sharp, P.A.: Splicing of messenger RNA precursors is inhibited by antisera to small nuclear ribonucleoprotein. Cell 35, 101-107 (1983)

Pelham, H.R.B., Brown, D.D.: A specific transcription factor that can bind either the 5S RNA gene or 5S RNA. Proc. Natl. Acad. Sci. USA 77, 4170-4174 (1980)

Pettersson, I., Hinterberger, M., Mimori, T., Gottlieb, E., Steitz, J.A.: The structure of mammalian small nuclear ribonucleoproteins. Identification of multiple protein components reactive with anti-(U_1) ribonucleoprotein and anti-Sm autoantibodies. J. Biol. Chem. 259, 5907-5914 (1984)

Picard, B., Wegnez, M.: Isolation of a 7S particle from Xenopus laevis oocytes. A 5S RNA-protein complex. Proc. Natl. Acad. Sci. USA 76, 241-245 (1979)

Piechaczyk, M., Lelay-Taha, M.N., Sri-Widada, J., Brunel, C., Liautard, J.P., Jeanteur, P.: Mouse DNA sequences complementary to small nuclear RNA U_1. Nucleic Acids Res. 10, 4627-4640 (1982)

Pizer, L.I., Deng, J.S., Stenberg, R.M., Tan, E.M.: Characterization of a phosphoprotein associated with the SS-B/La nuclear antigen in adenovirus-infected KB cells. Mol. Cell Biol. 3,1235-1245 (1983)

Pluskal, M.G., Sarkar, S.: Cytoplasmic low molecular weight ribonucleic acid species of chick embryonic muscles, a potent inhibitor of messenger ribonucleic acid translation in vitro. Biochemistry 20, 2048-2055 (1981)

Prestayko, A.W., Tonato, M., Busch, H.: Low molecular weight RNA associated with 28S nucleolar RNA. J. Mol. Biol. 47, 505-515 (1970)

Price, R., Penman, S.: Transcription of the adenovirus genome by an α-amanitin sensitive ribonucleic acid polymerase in HeLa cells. J. Virol. 9, 621-626 (1972)

Puvion, E., Viron, A., Assens, C., Leduc, E.H., Jeanteur, Ph.: Immuno-cytochemical identification of nuclear structures containing snRNPs in isolated rat liver cells. J. Ultrastr. Res. 87, 180-189 (1984)

Raj, N.B.K., Ro-Choi, T.S., Busch, H.: Nuclear ribonucleoprotein complexes containing U_1 and U_2 RNA. Biochemistry 14, 4380-4385 (1975)

Rao, D.D., Huang, A.S.: Synthesis of a small RNA in cells coinfected by standard and defective interfering particles of vesicular stomatitis virus. Proc. Natl. Acad. Sci. USA 76, 3742-3745 (1979)

Reddy, R., Busch, H.: U snRNAs of nuclear snRNPs. In: The Cell Nucleus (ed. H. Busch), Vol. VIII, pt. A, pp. 261-306. London-New York: Academic Press 1981

Reddy, R., Ro-Choi, J.S., Henning, D., Busch, H.: Primary sequence of U-1 nuclear ribonucleic acid of Novikoff Hepatoma ascite cells. J. Biol Chem. 249, 6486-6494 (1974a)

Reddy, R., Sitz, T.O., Ro-Choi, T.S., Busch, H.: Two-dimensional polyacrylamide gel electrophoresis separation of low molecular weight nuclear RNA. Biochem. Biophys. Res. Commun. 56, 1017-1022 (1974b)

Reddy, R., Henning, D., Busch, H.: Nucleotide sequence of nucleolar U_3 B RNA. J. Biol. Chem. 254, 11097-11105 (1979)

Reddy, R., Henning, D., Busch, H.: Substitutions, insertions and deletions in two highly conserved U_3 RNA species. J. Biol. Chem. 255, 7029-7033 (1980)

Reddy, R., Henning, D., Epstein, P., Busch, H.: Primary and secondary structure of U_2 snRNA. Nucleic Acids Res. 9, 5645-5658 (1981a)

Reddy, R., Li, W.Y., Henning, D., Choi, Y.C., Nohga, K., Busch, H.: Characterization and subcellular localization of 7-8S RNAs of Navikoff Hepatoma. J. Biol. Chem. 256, 8452-8457 (1981b)

Reddy, R., Henning, D., Busch, H.: The primary nucleotide sequence of U_4 RNA. J. Biol. Chem. 256, 3532-3538 (1981c)

Reddy, R., Spector, D., Henning, D., Liu, M.H., Busch, H.: Isolation and partial characterization of Dinoflagellate U_1-U_6 small RNAs, homologous to rat U small nuclear RNAs. J. Biol. Chem. 258, 13965-13969 (1983a)

Reddy, R., Tan, E.M., Henning, D., Nohga, K., Busch, H.: Detection of a nucleolar 7-2 ribonucleoprotein with autoantibodies from patients with scleroderma. J. Biol. Chem. 258, 1383-1386 (1983b)

Reed, R.E., Baer, M.F., Guerrier-Takada, C., Donis-Keller, H., Altman, S.: Nucleotide sequence of the gene encoding the RNA subunit (mRNA) of ribonuclease P from Escherichiae Coli. Cell 30, 627-636 (1982)

Reichel, R., Monstein, H.J., Jansen, H.W., Philipson, L., Benecke, B.J.: Small nuclear RNAs are encoded in the non transcribed region of ribosomal spacer DNA. Proc. Natl. Acad. Sci. USA 79, 3106-3110 (1982)

Rein, A.: The small molecular weight manodisperse nuclear RNA's in mitotic cells. Biochim. Biophys. Acta 232, 306-313 (1971)

Reveillaud, I., Lelay-Taha, M.N., Sri-Widada, J., Brunel, C.; Jeanteur, Ph.: Mg^{2+} induces a sharp and reversible transition in U_1 and U_2 snRNPs configuration. Mol. Cell. Biol. 4, 1890-1899 (1984)

Rinke, J., Steitz, J.A.: Precursor molecules of both human 5S ribosomal RNA and transfer RNAs are bound by a cellular protein reactive with anti-La lupus antibodies. Cell 29, 149-159 (1982)

Rinke, J., Appel, B., Blocker, H., Frank, R., Lührman, R.: The 5'-terminal sequence of U_1 RNA complementary to the consensus 5' splice site of premRNA is single-stranded in intact U_1 snRNP particles. Nucleic Acids Res. 12, 4111-4126 (1984)

Robert-Robin, J., Emanoil, Ravicovitch, R., Bazilier, M., Boiron, M.: Isolation and characterization of "8S" RNA in murine sarcoma virus-infected cells. Biochem. Biophys. Res. Commun. 60, 965-975 (1974)

Roi-Choi, T.S., Moriyama, Y., Choi, Y.C., Busch, H.: Isolation and purification of a nuclear 4.5S ribonucleic acid of the Novikoff Hepatoma. J. Biol. Chem. 245, 1970-1977 (1970)

Roi-Choi, T.S., Reddy, R., Henning, D., Takano, T., Taylor, C., Busch, H.: Nucleotide sequence of 4.5S ribonucleic acid of Novikoff hepatoma cell nuclei. J. Biol. Chem. 247, 3205-3222 (1972)

Ro-Choi, T.S., Raj, N.B.K., Pike, L.M., Busch, H.: Effects of α-amanitin, cycloheximide, and thioacetamide on low molecular weight nuclear RNA. Biochemistry 15, 3823-3828 (1976)

Rogers, J., Wall, R.: A mechanism for RNA splicing. Proc. Natl. Acad. Sci. USA 77, 1877-1879 (1980)

Roop, D., Kristo, P., Stumph, W.E., Tsai, M.J., O'Malley, B.W.: Structure and expression of a chicken gene coding for U_1 RNA. Cell 23, 671-680 (1981)

Rosa, M.D., Gottlieb, E., Lerner, M.R., Steitz, J.A.: Striking similarities are exhibited by two small Epstein-Barr virus-encoded ribonucleic acids and the adenovirus-associated ribonucleic acids VAI and VAII. Mol. Cell. Biol. 1, 785-796 (1981)

Rosa, M.D., Hendrick, J.P., Lerner, M.R., Steitz, J.A., Reichlin, M.: A mammalian $tRNA^{his}$ containing antigen is recognized by the polymyositis-specific antibody anti Jo-1. Nucleic Acids Res. 11, 853-870 (1983)

Saluz, H.P., Schmidt, T., Dudler, R., Altwegg, M., Stumm-Zollinger, E., Kubli, E., Chen, P.S.: The genes coding for 4 snRNAs of Drosophila melanogaster: localization and determination of gene numbers. Nucleic Acids Res. 11, 77-90 (1983)

Sass, H., Pederson, T.: Transcription-dependent localization of U_1 and U_2 small nuclear RNP's at major sites of gene activity in polytene chromosomes. J. Mol. Biol. 180, 911-926 (1984)

Savouret, J.F., Cathala, G., Eberhard, N.L., Miller, W.L., Baxter, J.D.: Interaction of small nuclear ribonucleoproteins with Simian Virus 40 in CV-1 cells: Is V_2 snRNA involved in regulating replication? DNA 3, 365-376 (1984)

Sawyer, R.C., Harada, F., Dahlberg, J.E.: Virion-associated RNA primer for Rous sarcoma virus DNA synthesis: Isolation from uninfected cells. J. Virol. 13, 1302-1311 (1974)

Schubert, M., Keene, J.D., Lazzarini, R.A., Emerson, S.V.: The complete sequence of a unique RNA species synthesized by a DI particle of VSV. Cell 15, 103-112 (1978)

Segall, J., Matsui, T., Roeder, R.G.: Multiple factors are required for the accurate transcription of purified genes by RNA polymerase III. J. Biol. Chem. 255, 11985-11991 (1980)

Setyono, B., Pederson, T.: Ribonucleoprotein organization of eukaryotic RNA. XXX, Evidence that U_1 small nuclear RNA is a ribonucleoprotein when base-paired with pre-messenger RNA in vivo. J. Mol. Biol. 174, 285-295 (1984)

Sharp, P.: Speculations on RNA splicing. Cell 23, 643-646 (1981)

Sharp, G.C., Irvin, W.S., Tan, E.M., Gould, R.G., Holman, H.R.: Mixed connective tissue disease. An apparently distinct rheumatic disease syndrome associated with a specific antibody to extractable nuclear antigen (ENA). Am. J. Med. 52, 148-159 (1972)

Shen, C.K.J., Maniatis, T.: The organization, structure and in vitro transcription of Alu family RNA polymerase III transcription units in the human-like globin gene cluster: precipitation of in vitro transcripts by lupus anti-La antibodies. J. Mol. Appl. Genet. 1, 343-360 (1982)

Shibata, H., Ro-Choi, T.S., Reddy, R., Choi, Y.C., Henning, D., Busch, H.: The primary nucleotide sequence of nuclear U_2 ribonucleic acid. The 5'-terminal portion of the molecule. J. Biol. Chem. 250, 3909-3920 (1975)

Sklar, V.E., Roeder, R.G.: Transcription of specific genes in isolated nuclei by exogenous RNA polymerases. Cell 10, 405-414 (1977)

Smith, J.H., Eliceiri, G.L.: Antibodies elicited by N^2, N^2-trimethyl guanosine react with small nuclear RNAs and ribonucleoproteins. J. Biol. Chem. 258, 4636-4640 (1983)

Soderlund, H., Pettersson, V., Vennstrom, B., Philipson, L., Mathews, M.B.: A new species of virus-coded low molecular weight RNA from cells infected with adenovirus type 2. Cell 7, 585-593 (1976)

Sri-Widada, J., Liautard, J.P., Assens, C., Brunel, C.: Primary structure identification of snRNAs present in highly purified snRNPs from HeLa cells. Mol. Biol. Rep. 8, 29-36 (1981)

Sri-Widada, J., Assens, C., Liautard, J.P., Jeanteur, P., Brunel, C.: Isolation of a pure U_1 snRNP from HeLa cells. Biochem. Biophys. Res. Commun. 104, 457-462 (1982)

Sri-Widada, J., Liautard, J.P., Brunel, C., Jeanteur, P.: Interaction of snRNA's with rapidly sedimenting nuclear sub-structures (hnRNP's) from HeLa cells. Nucleic Acids Res. 11, 6631-6646 (1983)

Stefano, J.E.: Purified lupus antigen La recognizes and oligouridylate stretch common to the 3' termini of RNA polymerase III transcripts. Cell 36, 145-154 (1984)

Steitz, J.A., Wolin, S.L., Rinke, J., Pettersson, I., Mount, S.M., Lerner, E.A., Hinterberger, M., Gottlieb, E.: Small ribonucleoproteins from eukaryotes: Structures and roles in RNA Biogenesis. Cold Spring Harbor Symp. Quant. Biol. 47, 893-900 (1983)

Stevenin, J., Gallinaro-Matringe, H., Gattoni, R., Jacob, M.: Complexing of the structure of particles containing heterogeneous nuclear RNA as demonstrated by ribonuclease treatment. Eur. J. Biochem. 74, 589-602 (1977)

Strub, K., Galli, G., Busslinger, M., Birnstiel, M.L.: The cDNA sequences of the sea urchin U_7 small nuclear RNA suggest specific contacts between histone mRNA precursor and U_7 RNA during RNA processing. EMBO J. 3, 2801-2807 (1984)

Stunnenberg, H.G., Birnstiel, M.L.: Bioassay for components regulating eukaryotic gene expression: a chromosomal factor involved in the generation of histone mRNA 3' termini. Proc. Natl. Acad. Sci. USA 79, 6201-6204 (1982)

Takano, M., Agris, P.F., Sharp, G.C.: Purification and biochemical characterization of nuclear ribonucleoprotein antigen using purified antibody from serum of a patient with mixed connective tissue disease. J. Clin. Invest. 65, 1449-1456 (1980)
Takano, M., Golden, S.S., Sharp, G.C., Agris, P.F.: Molecular relationships between two nuclear antigens ribonucleoprotein and Sm: purification of active antigens and their biochemical characterization. Biochemistry 21, 5929-5936 (1981)
Tamm, I., Kikuchi, T.: Early termination of heterogeneous nuclear RNA transcript in mammalian cells: accentuation by 5,6-dichloro-1-β-D-ribofuranosylbenzimidazole. Proc. Natl. Acad. Sci. USA 76, 5750-5754 (1979)
Tamm, I., Kikuchi, T., Darnell, J.E., Salditt-Georgieff, M.: Short capped hnRNA precursor chains in HeLa cells: continued synthesis in the presence of 5,6-dichloro-1-β-D-ribofuranosylbenzimidazole Biochemistry 19, 2743-2748 (1980)
Tan, E.M.: Autoimmunity to nuclear antigens. In: The Cell Nucleus (ed. H. Busch), Vol. VII, pp. 457-477. London-New York: Academic Press 1979
Tan, E.M.: Autoantibodies to nuclear antigens (ANA): their immunobiology and medicine. Adv. Immunol. 33, 167-240 (1982)
Tan, E.M., Kunkel, H.G.: Characteristics of a soluble nuclear antigen precipitating with sera of patients with systemic lupus erythematosus. J. Immunol. 96, 464-471 (1966)
Thimmappaya, B., Weinberger, C., Schneider, R.J., Shenk, T.: Adenovirus VAI RNA is required for efficient translation of viral mRNAs at late times after infection. Cell 31, 543-551 (1982)
Towbin, H., Staehlin, T., Gordon, J.: Electrophoretic transfer of proteins from acrylamide gels to nitrocellulose sheets. Procedure and some applications. Proc. Natl. Acad. Sci. USA 76, 4350-5354 (1979)
Udvardy, A., Seifart, K.N.: Transcription of specific genes in isolated nuclei from HeLa cells in vitro. Eur. J. Biochem. 62, 353-363 (1976)
Ullu, E., Melli, M.: Cloning and characterization of cDNA copies of the 7S RNAs of HeLa cells. Nucl. Acids. Res. 10, 2209-2223 (1982)
Ullu, E., Esposito, V., Melli, M.: Evolutionary conservation of the human 7S RNA sequences. J. Mol. Biol. 161, 195-201 (1982a)
Ullu, E., Murphy, S., Melli, M.: Human 7SL RNA consists of a 140 nucleotide middle-repetitive sequence inserted in an Alu sequence. Cell 29, 195-202 (1982b)
Van Arsdell, S.W., Weiner, A.M.: Human genes for U_2 small nuclear RNA are tandemly repeated. Mol. Cell. Biol. 4, 492-499 (1984a)
Van Arsdell, S.W., Weiner, A.M.: Pseudogenes for human U_2 small nuclear RNA do not have a fixed site of 3' truncation. Nucleic Acids Res. 12, 1463-1471 (1984b)
Van Arsdell, W.S., Denison, R.A., Bernstein, L.B., Weiner, A.M., Manser, T., Gesteland, R.F.: Direct repeats flank three small nuclear RNA pseudogenes in the human genome. Cell 26, 11-17 (1981)
Van Eekelen, C., Buijtels, H., Linne, T., Ohlsson, R., Philipson, L., Van Venrooij, W.: Detection of a cellular polypeptide associated with adenovirus-coded VA RNA using in vitro labeling of proteins cross-linked to RNA. Nucleic Acids Res. 10, 3039-3052 (1982)
Vogelstein, B., Hunt, B.F.: A subset of small nuclear ribonucleoprotein particle antigens is a component of the nuclear matrix. Biochem. Biophys. Res. Commun. 105, 1224-1232 (1982)
Wallace, R., Kates, J.: State of adenovirus 2 deoxyribonucleic acid in the nucleus and its mode of transcription: studies with isolated viral deoxyribonucleic acid-protein complexes and isolated nuclei. J. Virol. 9, 627-635 (1972)

Walter, P., Blobel, G.: Purification of a membrane-associated protein complex required for protein translocation across the endoplasmic reticulum. Proc. Natl. Acad. Sci. USA 77, 7112-7116 (1980)

Walter, P., Blobel, G.: Translocation of proteins across the endoplasmic reticulum II. Signal recognition protein (SRP) mediates the selective binding to microsomal membranes of in vitro-assembled polysomes synthesizing secretory protein. J. Cell Biol. 91, 551-556 (1981a)

Walter, P., Blobel, G.: Translocation of proteins across the endoplasmic reticulum III. Signal recognition on protein (SRP) causes signal sequence-dependent and site-specific arrest of chain elongation that is released by microsomal membranes. J. Cell Biol. 91, 557-561 (1981b)

Walter, P., Blobel, G.: Signal recognition particle contains a 7S RNA essential for protein translocation across the endoplasmic reticulum. Nature (London) 299, 691-698 (1982)

Walter, P., Ibrahimi, I., Blobel, G.: Translocation of protein across the endoplasmic reticulum I. Signal recognition protein (SRP) binds to in vitro assembled polysomes synthesizing secretory protein. J. Cell. Biol. 91, 545-550 (1981)

Watanabe-Nagasu, N., Itoh, Y., Tani, T., Okano, K., Koga, N., Okada, N., Ohshima, Y.: Structural analysis of gene loci for rat U_1 small nuclear RNA. Nucleic Acids Res. 11, 1791-1801 (1983)

Weill, P.A., Segall, J., Harris, B., Ng, S.Y., Roeder, R.G.: Faithful transcription of eukaryotic genes by RNA polymerase III in systems reconstituted with purified DNA templates. J. Biol. Chem. 254, 6163-6173 (1979)

Weimann, R., Brendler, T.G., Raskas, H.J., Roeder, R.G.: Low molecular weight viral RNAs transcribed by RNA polymerase III during adenovirus 2 infection. Cell 7, 557-566 (1976)

Weinberg, R.A., Penman, S.: Small molecular weight monodisperse nuclear RNA. J. Mol. Biol. 38, 289-304 (1968)

Weinberg, R., Penman, S.: Metabolism of small molecular weight monodisperse nuclear RNA. Biochim. Biophys. Acta 190, 10-29 (1969)

Weiner, A.M.: An abundant cytoplasmic 7S RNA is complementary to the dominant interspersed middle repetitive DNA sequence family in the human genome. Cell 22, 209-218 (1980)

Westin, C., Monstein, H.J., Zabielski, J., Philipson, L., Pettersson, U.: Human DNA sequences complementary to the small nuclear RNA U_2. Nucleic Acids Res. 9, 6323-6338 (1981)

Westin, G., Zabielski, J., Hammarstrom, K., Monstein, H.J., Bark, C., Pettersson, U.: Clustered genes for human U_3 RNA. Proc. Natl. Acad. Sci. USA 81, 3811-3815 (1984)

White, P.J., Hoch, S.O.: Definition of the antigenic polypeptides in the Sm and RNP ribonucleoprotein complexes. Biochem. Biophys. Res. Commun. 102, 365-371 (1981)

White, P.J., Gardner, W.D., Hoch, S.O.: Identification of the immunologically active components of the Sm and RNP antigens. Proc. Natl. Acad. Sci. USA 78, 626-630 (1981)

White, P.J., Billings, P.B., Hoch, S.O.: Assays for the Sm and RNP autoantigens: the requirement for RNA and influence of the tissue source. J. Immunol. 128, 2751-2756 (1982)

Wieben, E.D., Pederson, T.: Small nuclear ribonucleoproteins of Drosophila: Identification of U_1 RNA-associated proteins and their behavior during heat shock. Mol. Cell. Biol. 2, 914-920 (1982)

Wieben, E.D., Madore, S.J., Pederson, T.: Protein binding sites are conserved in small nuclear RNA U_1 from insects and mammals. Proc. Natl. Acad. Sci. USA 80, 1217-1220 (1983a)

Wieben, E.D., Madore, S., Pederson, T.: U_1 small nuclear ribonucleoprotein studied by in vitro assembly. J. Cell Biol. 96, 1751-1755 (1983b)
Wilk, H.E., Kecskemethy, N., Schafer, K.P.: m-aminophenylboronate agarose specifically binds capped snRNA and mRNA. Nucleic Acids Res. 10, 7621-7633 (1982)
Wise, J.A., Weiner, A.M.: Dictyostelium small nuclear RNA D2 is homologous to rat nucleolar RNA U_3 and is encoded by a dispersed multigene family. Cell 22, 109-118 (1980)
Wolin, S.L., Steitz, J.A.: Genes for two small cytoplasmic Ro RNA are adjacent and appear to be single copy in the human genome. Cell 32, 735-744 (1983)
Wooley, J.C., Cone, R.D., Tartof, D., Chung, S.Y.: Small nuclear ribonucleoprotein complexes of Drosophila melanogaster. Proc. Natl. Acad. Sci. USA 79, 6762-6766 (1982)
Wooley, J.C., Zukerberg, L.R., Chung, S.Y.: Polypeptide components of human small nuclear ribonucleoproteins. Proc. Natl. Acad. Sci. USA 80, 5208-5212 (1983)
Yang, V.W., Lerner, M.R., Steitz, J.A., Flint, S.J.: A small nuclear ribonucleoprotein is required for splicing of adenoviral early RNA sequences. Proc. Natl. Acad. Sci. USA 78, 1371-1375 (1981)
Zeller, R., Nyffenegger, T., De Robertis, E.M.: Nucleocytoplasmic distribution of snRNPs and stock piled snRNA-binding proteins during oogenesis and early development in Xenopus laevis. Cell 32, 425-434 (1983)
Zeller, R., Carri, M.T., Mattaj, I.W., De Robertis, E.M.: Xenopus laevis U_1 snRNA genes: characterization of transcriptionally active genes reveals major and minor repeated gene families. EMBO J. 3, 1075-1081 (1984)
Zieve, G.: Two groups of small stable RNAs. Cell 25, 296-297 (1981)
Zieve, G., Penman, S.: Small RNA species of the HeLa cell: Metabolism and subcellular localization. Cell 8, 19-31 (1976)
Zieve, G., Penman, S.: Subnuclear particles containing small nuclear RNA and heterogeneous nuclear RNA. J. Mol. Biol. 145, 501-523 (1981)
Zieve, G., Benecke, B.J., Penman, S.: Synthesis of two classes of small RNA species in vivo and in vitro. Biochemistry 16, 4520-4525 (1977)

Nucleotide-specific Ribonucleases from Eukaryotes. Their Possible Roles During Poly(A) (+)mRNA Maturation and Degradation

H. C. Schröder, M. Bachmann, R. Messer and W. E. G. Müller

A. Introduction

The primary gene transcripts of eukaryotes are large precursor RNA molecules, termed heterogeneous nuclear RNA (hnRNA), which must be posttranscriptionally modified by a series of modification steps to obtain the functional, cytoplasmic mRNA which is on an average three to five times smaller in size than the primary transcript (Darnell 1979). Several lines of evidence indicate that mRNA biosynthesis in eukaryotes is controlled not only at the level of DNA transcription but also at the level of posttranscriptional RNA processing (Darnell 1982; Perry et al. 1979). During processing of hnRNA to mRNA, a reduction of sequence complexity occurs (Wold et al. 1978). Nucleic acid hybridization experiments revealed a 20 times greater complexity for hnRNA than for mRNA (Chikaraishi et al. 1978). Therefore, it was concluded (Darnell 1979) that the number of different kinds of hnRNA's originally synthesized is five times greater than that being processed to mRNA. Ono and Cutler (1978) presented evidence suggesting that in differentiated cells some genes are transcribed from which the translation products never appear in the cytoplasm; the amount of these "heterologous" hnRNA increases during the ageing of the animal. Moreover, albumin mRNA precursors were found to be present in the nuclei of liver cells of analbuminemic rats lacking the corresponding mRNA within their cytoplasm (Esumi et al. 1982). Comparing the sequence complexity of the transcribed unique sequence DNA with that of cytoplasmic mRNA, Imaizumi-Scherrer et al. (1982) even concluded that control of expression of the duck genome occurs predominantly at the posttranscriptional level. On the other hand, Meyuhas and Perry (1979) suggested that in the case of very abundant mRNA's the determination of mRNA frequency distributions takes place primarily on the transcriptional level, whereas in the case of sparse mRNA's posttranscriptional control mechanisms predominate. In avian erythroblastosis virus transformed cells globin mRNA sequences were found to be restricted exclusively to the nucleus indicating a posttranscriptional suppression of globin gene expression within these cells (Therwath and Scherrer 1978).

Despite these numerous findings supporting the "cascade regulation scheme" (Imaizumi-Scherrer et al. 1982) of gene expression, only little is known at which steps of posttranscriptional processing (Sect. B) control occurs.

Recently Milner et al. (1984) demonstrated the presence of "identifier sequences" within the introns of rat brain-specific hnRNA precursors, possibly involved in posttranscriptional control of brain-specific gene expression. Moreover, some cases of alternative splicing are known resulting in the formation of different mRNA's encoding different types of proteins from one gene, e.g., in the course of expression of the calcitonin gene (Amara et al. 1982), the murine αA crystallin gene (King and Piatigorsky 1983) and the chicken myosin light chain gene (Nabeshima et al. 1984). In addition, different processing times may also contribute to the regulation of the level of individual cytoplasmic mRNA's as shown in *Dictyostelium discoideum* aggregates (Mangiarotti et al. 1983). Moreover, the bulk of the transcripts of the *Drosophila* reiterated *copia* sequences were found to accumulate within the nucleus where it becomes degraded, although the corresponding cytoplasmic poly(A)(+)mRNA's are quite stable (Falkenthal et al. 1982).

B. Posttranscriptional Events in mRNA Maturation

Processing of hnRNA proceeds by a series of modification steps, during which both degradation of primarily transcribed sequences and synthesis of novel nucleotide stretches takes place. Twenty-five to thirty nucleotides upstream from the TATAAT sequence, RNA polymerase II begins to transcribe the sequences of the primary transcript encoded in the transcriptional unit of DNA (Nevins 1983); see Fig. 1. The synthesis of the primary transcript does not terminate at the poly(A) addition site but proceeds far beyond this sequence. Most hnRNA precursors are characterized by the presence of noncoding, intervening sequences (introns) which are interspersed between the coding RNA sequences (exons). The first step in hnRNA processing consists in the formation of the 5'-terminal cap structure (Shatkin 1976). After a cleavage process at the 3' end of the primary transcript (Nevins and Darnell 1978) which occurs about twenty base pairs upstream from a AAUAAA consensus sequence (Proudfoot and Brownlee 1976), a poly(A) sequence of up to 225 AMP units is synthesized at the newly created 3'-OH terminus of hnRNA (Molloy et al. 1972). The formation of the 3'poly(A) tail of hnRNA occurs by stepwise addition of AMP residues from ATP catalyzed by a poly(A) polymerase in a template-independent manner (Edmonds 1982). Thereafter, a few adenosine nucleotides within the transcribed part of mRNA (Perry et al. 1974) as well as within the poly(A) tract (Nichols and Welder 1981) become modified by methylation. During the RNA:RNA splicing

---

Fig. 1. Posttranscriptional processing steps in poly(A)(+)mRNA biosynthesis. ► *Black bars* coding sequences; *hatched bars* noncoding sequences; *thin lines* sequences excised and degraded during processing; *X* poly(A) initiation site; *l* leader sequence. (References: Roeder and Rutter 1980; Shatkin 1976; Darnell 1982; Proudfoot and Brownlee 1976; Manley 1983a,b; Edmonds and Abrams 1960; Blobel 1973; Rothman et al. 1974; Lerner et al. 1980; Jeffreys and Flavell 1977; Bachmann et al. 1983a,b, 1984b; Cech 1983; Darnell et al. 1973; Agutter et al. 1976; Bernd et al. 1982b; Müller 1976; Bergmann and Brawerman 1977; Schröder et al. 1980a,b)

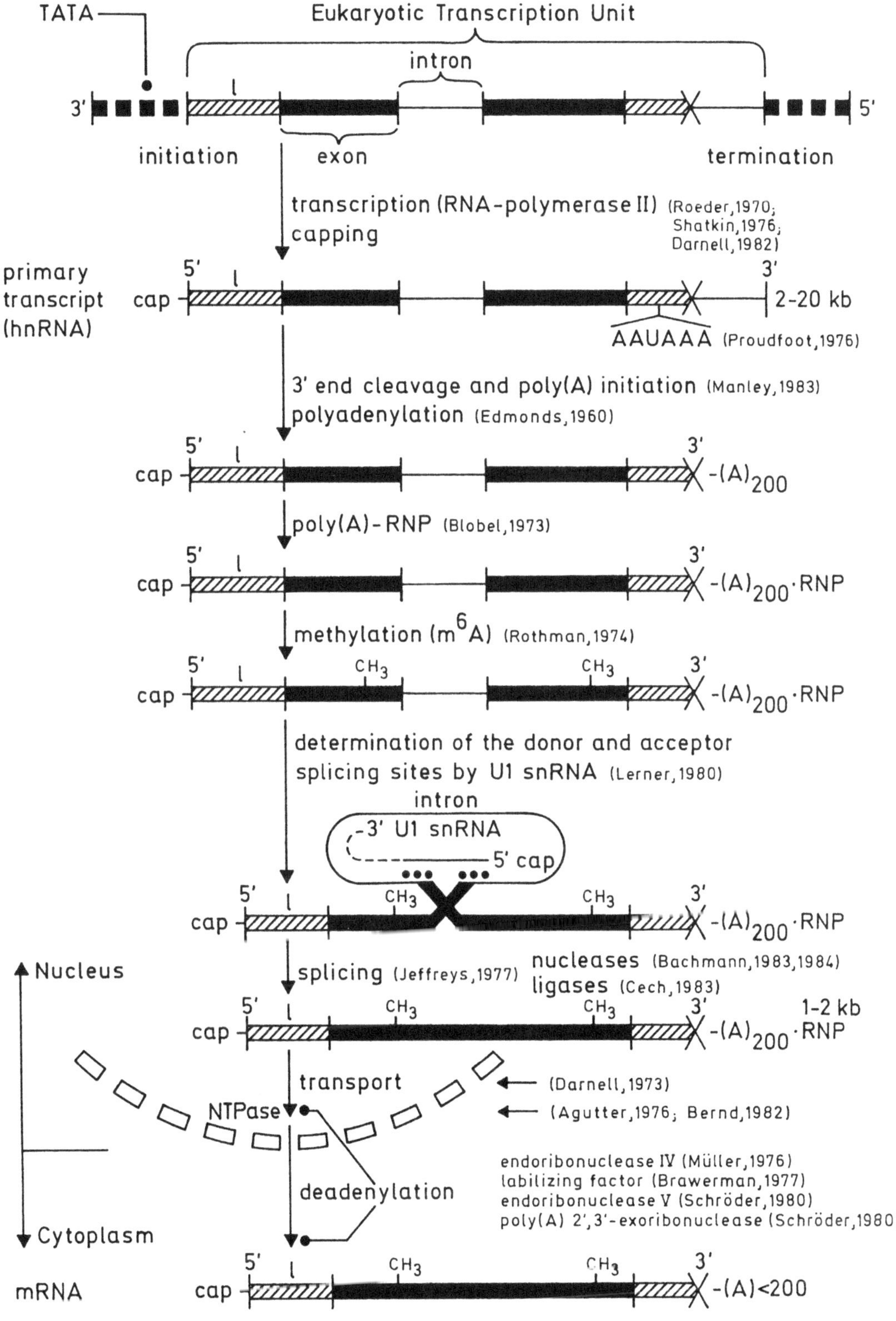
TATA
Eukaryotic Transcription Unit
intron
3'
5'
initiation
exon
termination
transcription (RNA-polymerase II) (Roeder,1970; Shatkin,1976; Darnell,1982)
capping
primary transcript (hnRNA)
cap
2-20 kb
AAUAAA (Proudfoot,1976)
3' end cleavage and poly(A) initiation (Manley,1983)
polyadenylation (Edmonds,1960)
-(A)200
poly(A)-RNP (Blobel,1973)
-(A)200·RNP
methylation (m6A) (Rothman,1974)
CH3
determination of the donor and acceptor splicing sites by U1 snRNA (Lerner,1980)
intron
-3' U1 snRNA
5' cap
Nucleus
splicing (Jeffreys,1977)
nucleases (Bachmann,1983,1984)
ligases (Cech,1983)
1-2 kb
transport
NTPase
(Darnell,1973)
(Agutter,1976; Bernd,1982)
deadenylation
endoribonuclease IV (Müller,1976)
labilizing factor (Brawerman,1977)
endoribonuclease V (Schröder,1980)
poly(A) 2',3'-exoribonuclease (Schröder,1980)
Cytoplasm
mRNA
-(A)<200

process (Jeffreys and Flavell 1977) the intron sequences within the hnRNA precursor are sequentially excised and degraded, and the thereby created open ends of the exons are spliced together by a yet unknown RNA ligase. Soon after transcription the hnRNA precursors become associated with specific proteins under formation of 200 - 300S hnRNP particles (Samarina and Krichevskaya 1981; Samarina et al. 1968). The subunits of these large complexes, called informofers, sedimenting at 30S, consist of six proteins of M_r 32,500 - 42,500 interacting with one another under formation of a protein core; the hnRNA is wound around the surface of these particles (Economidis and Pederson 1983). Moreover, the poly(A) tail of hnRNA as well as mRNA forms a complex with specific poly(A)-associated proteins (Blobel 1973). During ageing of poly(A)(+)mRNA the length of the poly(A) tail is reduced both in the nucleus and in the cytoplasm (Sheiness and Darnell 1973; Brawerman 1973) by the action of poly(A)-specific ribonucleases (Müller 1976; Schröder et al. 1980a,b); however, poly(A) has also been shown to be newly synthesized in the cytoplasm (Edmonds and Winters 1976). With some exceptions, e.g. histone mRNA's, the deadenylated nuclear mRNA's are not transported to the cytoplasm (Weinberg 1973).

Among the enzymes involved in poly(A)(+)mRNA maturation, ribonucleases must be predicted to play a prominent role. In this review we summarize the properties primarily of those enzymes which are involved in poly(A) catabolism (Sect. D) or potential function during the hnRNA splicing process (Sect. E).

I. Capping

The earliest posttranscriptional modification step is the formation of the 5'terminal cap structure of mRNA (Shatkin 1976), which seems to proceed even before the primary transcript is completely synthesized (Babich et al. 1980). This structure in whose synthesis at least four different enzymes are involved (Perry 1981) consists of a 7-methylguanosine residue (m^7G) which is linked by a 5'-5' triphosphate bridge with the penultimate nucleotide. The ribose moiety of the latter nucleotide and in some cases also that of the third nucleotide can be methylated, so that three types of cap structures can be distinguished: cap 0, $m^7G(5')ppp(5')XpYpZp$; cap 1, $m^7G(5')ppp(5')X^mpYpZp$, and cap 2, $m^7G(5')ppp(5')X^mpY^mpZp$. The m^7G of the 5'cap seems to be essential for the formation of the initiation complex of protein synthesis (Rhoads 1985); the adjacent 5'terminal leader sequence has been proposed to support the mRNA-ribosome interaction by base-pairing with the 3'end of 18S rRNA (Nakashima et al. 1980). In addition, the 5'cap may function in protecting the mRNA against exonucleolytic degradation from the 5'terminus (Nevins 1983).

II. Methylation

The function of the hnRNA methylation which proceeds at the N^6 position of internal adenylate residues (Perry et al. 1974) is not yet clear. It was proposed that the internal m^6A residues which are conserved during processing (Chen-Kiang et al. 1979) may be involved in splicing or required to prevent endonucleolytic cleavage of the mRNA (Stoltzfus and Dane 1982).

III. Polyadenylation

Polyadenylate sequences with chain lengths up to 250 AMP units have been found at the 3'terminus of most eukaryotic and viral mRNA's (Littauer and Soreq 1982). 3'Poly(A) or 3'oligo(A) segments are not restricted to eukaryotes, as assumed previously (Brawerman 1974), but have also been detected in prokaryotic organisms, although they are shorter in size (Carlin 1978). The base composition of the poly(A) segment must not be strictly uniform since it can be interspersed by a few modified nucleotides which are most likely 2'-O-methyl-AMP's (Nichols and Welder 1981). Although the bulk of eukaryotic mRNA's appears to be polyadenylated, there are important exceptions. Thus, most histone mRNA's lack poly(A). Moreover, a highly complex population of poly(A)(-)mRNA could be detected in the brain of adult animals (Chaudhari and Hahn 1983). However, it cannot be excluded that histone mRNA's could also be initially polyadenylated follwed by a rapid deadenylation, since some histone mRNA's contain poly(A) sequences (Levenson and Marcu 1976; Ruderman and Pardue 1977).

Several functions have been attributed to the 3'terminal poly(A) sequence, but only a few are experimentally well proven. The polyadenylation of hnRNA and mRNA may be involved in (1) stabilizing of the RNA molecule (Nudel et al. 1976); (2) enabling nucleocytoplasmic mRNA translocation (Bernd et al. 1982b; Müller et al. 1983, 1984); (3) determining the translational capacity of mRNA (Jacobsen and Favreau 1983); (4) facilitating mRNA splicing (Bina et al. 1980); and (5) terminating of the reading frames in polycistronic RNA molecules (Ojala et al. 1981).

Microinjection experiments with mRNA fractions containing poly(A) of different sizes using *Xenopus* oocytes support the assumption that poly(A) has a stabilizing role (Nudel et al. 1976). However, below a critical chain length of 10 to 20 AMP units, mRNA stability could not be assured any more. Moreover, it was found that no proportionality exists between stability and the size of poly(A). The 3'terminal poly(A) segment may act in preventing the degradation of the mRNA by those ribonucleases that are not able to hydrolyze poly(A) or are inhibited by poly(A) (Levy et al. 1975).

In order to determine whether poly(A) is required for translation of mRNA, globin mRNA (Sippel et al. 1974; Soreq et al. 1974), total mRNA from mouse L cells (Bard et al. 1974) and quail oviduct (Müller et al. 1976) were deadenylated by different specific procedures. Since thereby the initial rates of translation of polyadenylated and polydeadenylated mRNA were found to be identical, it was concluded that poly(A) has no effect on the translational capacity of the mRNA. Recently, however, it could be documented by competition experiments (Jacobson and Favreau 1983) that exogenous poly(A) inhibits mRNA translation raising once more the question whether 3'poly(A) is involved in protein synthesis.

The functions of the 3'terminal poly(A) sequence in mRNA splicing and nucleocytoplasmic mRNA transport will be discussed in Sections E.II and F.

IV. Splicing

The primary transcripts of most eukaryotic genes contain besides exon regions, which are conserved during processing to mRNA, intron sequences which are removed during the splicing reactions within the nucleus (Abelson 1979). It is assumed that the hnRNA splicing usually follows poly(A) addition, although it has not been clearly proven (Darnell 1982). Compared to the polyadenylation process which is completed within a few minutes (Weber et al. 1980), splicing of hnRNA is slow and requires approx. 20 min (Nevins 1979). The mechanism by which splicing occurs is not yet fully understood. However, several lines of evidence indicate small nuclear RNP's (snRNP's) are involved in splicing (Padgett et al. 1983; Lerner et al. 1980). The function of U1 snRNA in hnRNA splicing will be discussed in Section E. Moreover, the 3'terminal poly(A) sequence of hnRNA has been proposed to facilitate the splicing process by promoting the formation of triple-stranded structures within the hnRNA to be spliced (Bina et al. 1980). In two recent contributions the occurrence of branched intermediates of hnRNA containing 2',5'-phosphodiester bonds (Wallace and Edmonds 1983) and the presence of 2-5A and of 2-5A metabolizing enzymes in nuclei of HeLa cells not treated with interferon (Nilsen et al. 1982b) has been reported, indicating a possible involvement of the 2-5A system in hnRNA splicing (Nilsen et al. 1982a).

V. Association with the Nuclear Matrix and the Cytoskeleton

The hnRNA and mRNA have been described to be bound to the nuclear matrix (T.E. Miller et al. 1978; Herman et al. 1978) and cytoplasmic mRNA and polyribosomes associated with the cytoskeleton (Lenk et al. 1977; Van Venrooij et al. 1981). The association of hnRNA with the nuclear matrix seems to take place already during transcription as suggested by Jackson et al. (1981). Recently Ciejek et al. (1983) reported that only the transcribed regions of actively expressed genes are associated with the matrix. The hnRNA appears to remain matrix-associated throughout processing (Long et al. 1979). It was proposed that the poly(A) sequence and the double-stranded regions serve as attachment points for the hnRNA to the nuclear matrix (Herman et al. 1978). In UV irradiated cells Van Eekelen and Van Venrooij (1981) identified two major hnRNP's with M_r = 41,500 and M_r = 43,000 to be cross-linked to hnRNA suggesting their involvement in hnRNA binding to the nuclear matrix.

VI. Nucleocytoplasmic mRNA Transport

Poly(A)(+)mRNA is transported from nuclei to cytoplasm through the nuclear pore complexes as poly(A)(+)mRNP complex (Webb et al. 1981). It is well established that RNA efflux from nuclei is an energy-dependent process (Ishikawa et al. 1978). Agutter et al. (1976, 1979) presented convincing evidence that a nuclear envelope nucleoside triphosphatase (NTPase) is involved in nucleocytoplasmic poly(A)(+)mRNA translocation. This enzyme was shown to be stimulated by the poly(A) segment of mRNA (Bernd et al. 1982b). The role of the 3'poly(A) tail of mRNA in modulating nuclear envelope NTPase activity will be discussed in Section F.

C. Eukaryotic Ribonucleases: A Brief Survey

Several ribonucleases with more or less nucleotide-specificity have been isolated from eukaryotic tissues. However, from most of them the physiological function is not yet known. Ribonucleases can be subdivided into exoribonucleases producing exclusively mononucleotides and endoribonucleases cleaving the polynucleotide substrate at internal positions. Also the endo-exoribonucleases which degrade their polynucleotide substrate under intermediate formation of oligomers up to monomers belong to the latter group.

The properties of both prokaryotic and eukaryotic nucleases have been excellently reviewed by Sierakowska and Shugar in 1977. More recently, Altman et al. (1982) have summarized the current state of knowledge about prokaryotic and eukaryotic nucleases involved in RNA processing. The following is a short description of the well known ribonucleases from eukaryotes and their main characteristics.

Endoribonuclease I (EC 3.1.4.22) isolated from pancreas, degrades preferentially poly(C) and poly(U) to P-2':3' nucleosides; the enzyme requires Mg^{2+} for activity, has two pH optima of pH 4.5 and 7.5, and an M_r of 14,000 (for references see Sierakowska and Shugar 1977). The acid endoribonuclease II (EC 3.1.4.23) which has been purified from several mammalian tissues displays no base specificity and produces oligonucleotides with 3'terminal 2':3'-cyclic phosphates; it has an M_r of 26,000 (13,000), a pH optimum of 5 to 6 and is Mg^{2+}-independent (Sierakowska and Shugar 1977; Bernardi and Bernardi 1966). An endoribonuclease II-like acid endoribonuclease has been purified from beef brain nuclei (Ittel et al. 1975); this enzyme, which has a pH optimum of 6.2, hydrolyzes poly(A), poly(U), and poly(C), and produces 3'-AMP from poly(A) as substrate. Endoribonuclease III (EC 3.1.4.24) has been originally found in *E. coli* (Robertson et al. 1968), a neutral endonuclease with similar properties has been later isolated from calf thymus (Ohtsuki et al. 1977); it cleaves double-stranded RNA under formation of oligonucleotides with 5'-P and 3'-OH termini; further characteristics are pH optimum 7.5, M_r of 60,000, and Mg^{2+}-dependence. The poly(A)-specific endoribonuclease IV (EC 3.1.26.6) which has been isolated and purified to homogeneity from chick oviduct is characterized by a pH optimum of 8.7, an

M_r of 45,000, the dependence on divalent cations, and the cleavage of poly(A) to oligo(A)$_{10}$ fragments containing 3'-OH and 5-P termini (Müller 1976, Müller et al. 1976). The main properties of the poly(A) and poly(U)-specific endoribonuclease V (EC 3.1.27.8) which has been purified to homogeneity from calf thymus gland (Schröder et al. 1980a) are M_r of 52,300, optimum pH of 5.8, and endo-exonucleolytic degradation of poly(A) up to 3'-AMP. The poly(U) and poly(A)-specific endoribonuclease VI from *Artemia larvae* (Quintanilla and Renart 1982) forms P-2':3'-Urd and 3'-UMP from poly(U) as substrate and exhibits high specificity toward UpN(3' → 5') phosphodiester linkages; it requires Ca^{2+} for activity and has an M_r of 38,000 and a pH optimum of 7.5. The poly(A)-modulated and oligo(U)-recognizing endoribonuclease VII (Bachmann et al. 1983b) is specific for single-stranded ribopyrimidines and produces oligoribonucleotides of an average chain length of 12 nucleotide units which are terminated by 3'-OH and 5'-P groups. A poly(C)-specific endoribonuclease has been isolated from chicken liver (Levy and Karpetsky 1980), this enzyme, which has an optimum pH between 6.5 and 7.5, is inhibited by synthetic poly(A) as well as the poly(A) tract of natural poly(A)(+) mRNA. A poly(A) and poly(U)-degrading endoribonuclease, which binds deoxyribonucleoside and ribonucleoside triphosphates and also cleaves denatured DNA, has been isolated from rat liver endoplasmic reticulum (Kouidou et al. 1981). A microsomal endoribonuclease, which displays very similar properties to both endoribonucleases IV and VI, has been purified from rat liver (Kumagai et al. 1980); the main characteristics are: M_r of 56,000, degradation of poly(A) and poly(U) to oligonucleotides with 3 to 7 residues in length, requirement for divalent cations (Mg^{2+} or Mn^{2+}) and inhibition by monovalent cations. A nuclear endoribonuclease that degrades poly(C) and poly(U) as well as rRNA and hnRNA has been isolated from HeLa cells, where it is predominantly localized in the nucleoli (Kwan 1976). Another nucleolar endoribonuclease that degrades poly(A), poly(C), and poly(U) has been purified from Ehrlich ascites tumor cells (Eichler and Eales 1982). From human liver a ribonuclease has been purified that does not split synthetic homopolynucleotides but synthetic copolymers containing AMP residues (Frank and Levy 1976); the activity of this enzyme, which produces oligonucleotides with 4 to 11 nucleotide residues, was shown to be inhibited by the poly(A) sequence of poly(A)(+)mRNA. The pyrimidine-specific ribonuclease from bovine seminal plasma (E.S.P. Reddy et al. 1979) which is a highly basic protein (pI, 9.8) degrades single- and double-stranded polyribonucleotides to 3'(2')-nucleotides; only Mg^{2+}-depleted, naturally occurring RNA's (rRNA and tRNA) are hydrolyzed by the enzyme, whereas the Mg^{2+}-containing RNA's are cleaved only to a very limited extent. A well-characterized endoribonuclease, which has been isolated from eukaryotic cells, is the hybrid nuclease (RNase H; EC 3.1.4.34); this nuclease specifically cleaves the RNA moiety of RNA:DNA hybrids (Hausen and Stein 1970; H.I. Miller et al. 1973). Another very interesting ribonuclease is the ppp(A2'p)$_n$A-dependent endoribonuclease (RNase L or F; for review see Sen 1982). This enzyme, which could be purified from different mammalian sources (Slattery et al. 1979; Wreschner et al. 1982), hydrolyzes in the presence of divalent cations only poly(U) (Slattery et al. 1979; Wreschner et al. 1981);

single-stranded RNA is cleaved primarily after AU, UG, and UU sequences (Floyd-Smith et al. 1981; Wreschner et al. 1981).

Some ribonucleases involved in maturation of RNA have been isolated from prokaryotes, but very little is known about eukaryotic processing nucleases. The most prominent enzyme of this group is ribonuclease P from *E. coli*, which cleaves a tRNA precursor (Kole and Altman 1981; Stark et al. 1978); another example is the membrane-associated endoribonuclease from yeast which also cuts pre-tRNA (Peebles et al. 1983). Moreover, endoribonuclease III from *E. coli* has been shown to process both *E. coli* 30S rRNA and bacteriophage T7 early RNA precursor (Robertson et al. 1968; Dunn and Studier 1973). Ribonuclease P-like activities have also been purified from eukaryotic sources (Altman et al. 1982). The ribonuclease NU, which cleaves a tRNA precursor from *E. coli*, has been isolated from human KB cells (Bothwell and Altman 1975).

The following ribonucleases are found to be particle-bound: an hnRNP-associated nonspecific endoribonuclease described by Niessing and Sekeris (1970); an hnRNP-associated endoribonuclease specific for double-stranded RNA described by Rech et al. (1979); the 45S RNP particle-bound endoribonuclease VII (Bachmann et al. 1983b; see above); the 12S RNP particle-associated, poly(U) and poly(C)-degrading pI 4.1 endoribonuclease (Bachmann et al. 1984a) and the U1 snRNP-associated poly(U)-specific endoribonuclease (Bachmann et al. 1983a).

Besides endoribonucleases and endo-exoribonucleases, some enzymes which act exclusively exonucleolytically have been purified from eukaryotic cells. An exoribonuclease (EC 3.1.4.20), which degrades single-stranded RNA to nucleoside 5'-monophosphates, has been purified from rat liver nuclei and Ehrlich ascites tumor cell nuclei (Lazarus and Sporn 1967; Sporn et al. 1969); the enzyme which is active in the presence of divalent cations (Mg^{2+} or Mn^{2+}) and an alkaline pH, displays some preference for poly(A), but also hydrolyzes poly(I), poly(U), and poly(C). The nucleolytic attack starts at the free 3'-OH end of the polynucleotide and is processive in nature, i.e., one RNA chain remains bound to the enzyme until it is completely hydrolyzed. A nuclear poly(A) specific 2',3'-exoribonuclease, which selectively degrades the 3'-poly(A) tail of hnRNA or mRNA to 5'-AMP's, has been isolated from different eukaryotic cells and purified to homogeneity from calf thymus gland (Schröder et al. 1980b). This enzyme is not only specific for oligo(A) with 3',5'-internucleotide bonds, but it also cleaves the 2',5'-linked oligomers (2-5A; Müller et al. 1980a). A cytoplasmic exoribonuclease, which also produces 5'-AMP, has been isolated from HeLa cells (Kwan 1977); this enzyme, which exhibits no base specificity, requires a free 3'-OH terminus of the polynucleotide and acts in a 3'→5 direction. A poly(A), poly(U) and poly(C)-degrading exoribonuclease purified from bovine brain (Guha 1975, 1977) cleaves polyribonucleotides only from their unphosphorylated 5'-OH ends to create 3'-mononucleotides; it has an M_r of 65,000, an optimum pH of 6 and requires no divalent cations for activity.

In the following Sections, the purification procedures, properties, and presumed functions of the known poly(A) catabolic enzymes (endoribonuclease IV, endoribonuclease V, and poly(A) specific 2',3'-exoribonuclease) as well as those of the pyrimidine-specific endoribonucleases associated with snRNP-hnRNP complexes (endoribonuclease VII, U1 snRNP-associated poly(U)-specific endoribonuclease, and 12S snRNP-associated pI 4.1 endoribonuclease) are described.

D. Poly(A) Metabolism

The chain length of the 3'terminal poly(A) sequences of hnRNA and mRNA is not transcriptionally controlled by a corresponding poly(dT) template on DNA (with some exceptions as Vaccinia virus RNA, Kates and Beeson 1970) but is posttranscriptionally regulated by a finely tuned equilibrium between polyadenylation and polydeadenylation events occurring in both the nuclear and cytoplasmic compartment (as well as in the mitochondrion in the case of mitochondrial mRNA) (Sheiness and Darnell 1973; Littauer and Soreq 1982; Müller et al. 1983). During ageing of an individual mRNA species, a reduction of the initial poly(A) size from around 250 AMP residues to less than 50 nucleotides takes place resulting in a heterogeneous poly(A) chain length distribution of the overall mRNA population within the cell. The chain length of poly(A) in a steady state is the result from both the defined equilibrium between poly(A) anabolic and poly(A) catabolic enzymes (coarse adjustment) and their modulation by posttranslational modification or interaction with certain poly(A)-associated proteins (fine tuning; see Sect. III). In addition, this equilibrium depends on a series of physiological factors as development, hormonal status, and nutrition (see Sect. IV).

I. Poly(A) Anabolic Enzymes

The poly(A) synthesis is catalyzed by poly(A) polymerase(s) and occurs at the 3'terminus of hnRNA 10 to 30 nucleotides downstream from the AAUAAA consensus sequence (Proudfoot and Brownlee 1976). In contrast to poly(A) catabolic nucleases, poly(A) polymerases have already been known for over 20 years (Edmonds and Abrams 1960), although some aspects of its action as its functional interaction with a specific endoribonuclease during RNA cleavage/poly(A) initiation is presently not yet understood (see next chapter).

1. Initiation of Polyadenylation and RNA Cleavage

It is well established that DNA transcription proceeds beyond the poly(A) initiation site (Darnell 1982). Therefore, a nucleolytic cleavage step is inserted before polyadenylation can start. The 3'terminal segment of the primary transcript split off by this process has not yet been seen; perhaps when it is formed, it is immediately degraded from its uncapped 5'end (Green et al. 1983). Although an exonucleolytic degradation up to the poly(A) start site cannot be ruled out, a specific endonuclease, perhaps complexed with the poly(A) polymerase, has been implicated in 3'end formation (Manley 1983a,b). The AAUAAA consensus sequence occurring 10 - 30 nucleotides upstream from the 3'-OH poly(A) priming site of hnRNA has been proposed to be a recognition signal for poly(A) polymerizing system (Proudfoot and Brownlee 1976). However, this sequence does not seem to be essential for the polyadenylation process per se. Thus, a point mutation resulting in a single base change (U → G) in the AAUAAA sequence of early region 1A mRNA of adenovirus has been shown to decrease the efficiency of RNA cleavage but not the efficiency of polyadenylation of those transcripts which are cleaved (Montell et al. 1983). The conclusion that AAUAAA is required for RNA cleavage, but not for polyadenylation is supported by the recent findings (Higgs et al. 1983) that patients suffering from a nondeletion type of α-thalassaemia which possess a single U → G transversion on the poly(A) initiation sequence of their α2 gene transcripts produce α-globin mRNA's with uncleaved 3'termini. Therefore, it is assumed (Montell et al. 1983; Manley 1983a,b) that RNA cleavage and poly(A) initiation are tightly coupled events which are possibly catalyzed by a complex between an until yet unknown endoribonuclease and a poly(A) polymerase. In a recent contribution Berget (1984) proposed U4 snRNA's to be involved in cleavage by base-pairing with the AAUAAA consensus sequence and a CAPyUG (Py = pyrimidine) sequence occuring close to the 3'OH polyadenylation site (Benoist et al. 1980). The assumption that sequences other than AAUAAA are signaling for the putative endoribonuclease·poly(A) polymerase complex, is supported by the finding that, in spite of identically positioned AAUAAA sequences, RNA's are found to be microheterogeneous with respect to the choice of the site of poly(A) addition (Fitzgerald and Shenk 1981; Sasavage et al. 1982). In contrast to the hitherto existing opinion that 3'end formation of histone mRNA's, which lack the AAUAAA sequence, occurs by simple transcription termination it seems to be clear now that at least some histone mRNA's create their 3'ends by posttranscriptional cleavage (Krieg and Melton 1984) and – and this will be important – with the aid of an snRNP particle (Galli et al. 1983). Alike the AAUAAA sequence in nonhistone mRNA's, a highly conserved sequence of 23 nucleotides is found at the 3'termini of histone mRNA's (Hentschel and Birnstiel 1981) which appears to be necessary for generation of correct 3'ends (Birchmeier et al. 1982).

2. Poly(A) Polymerases

The synthesis of the 3'terminal poly(A) segment of hnRNA and mRNA is catalyzed by poly(A) polymerase(s) (EC 2.7.7.19) which belong(s) to the group of primer-dependent and template-independent nucleotide polymerizing enzymes (Edmonds and Winters 1976; Edmonds 1982). Poly(A) polymerases have been found in nuclei (Winters and Edmonds 1973a,b; Müller et al. 1975, 1977a; Schröder et al. 1984c), cytoplasm (Tsiapalis et al. 1975) and mitochondria (Rose et al. 1975). Nuclear poly(A) polymerase has been reported to be present firstly in a soluble, nucleoplasmic state (Rose et al. 1977) and secondly bound to nuclear substructures as chromatin (Rose et al. 1977) or nuclear matrix (Schröder et al. 1984c). The mol. wt. of the purified enzymes varied between 48,000 in rat liver (Rose and Jacob 1976) and 140,000 in

Table 1. Properties of the known poly(A) catabolic enzymes, endoribonuclease IV (EC 3.1.26.6), endoribonuclease V (EC 3.1.27.8) and 2',3'-exoribonuclease (EC 3.1.13.4); n.d. = not determined

	Endoribonuclease IV	Endoribonuclease V	2',3'-exoribonuclease
Source	Chick oviduct	Calf thymus	Calf thymus
Purification	Homogeneous (15,000-fold)	Homogeneous	Homogeneous (8300-fold)
Molecular weight	45,000	52,300	80,000 (58,000 + 31,000)
pH optimum	8.7	5.8	9.0
Isoelectric point	7.2	6.3	n.d.
Temperature optimum	37°C	37 - 38°C	39°C
Cofactor	Mn^{++} and Mg^{++}	None	Mg^{++} and Mn^{++}
Specificity	Poly(A)	Poly(A) and poly(U)	Poly(A) and poly(A) • poly(U)
Product	Oligo$(pA)_{\overline{10}}$	3'-AMP(3'-UMP)	5'-AMP
Cleavage rate (cleavages/molecule × min)	1.7×10^3	3.7×10^3	1.2×10^2
Concentration (molecules/cell)	4.0×10^4	1.6×10^5	1.6×10^2
Localization	Nucleus (cytoplasm?)	n.d.	Nucleus

calf thymus (Winters and Edmonds 1973a,b). In view of the high affinity of poly(A) polymerase to the poly(A)-specific endoribonuclease IV (see Sect. D.II.2.c) it is not always clear if all the preparations of the enzyme are free from poly(A) degrading activity (Abraham and Jacob 1978) and if not, some of them contain the poly(A) polymerase-endoribonuclease IV complex. All poly(A) polymerases tested so far displayed very little specificity for the base sequence of the initiator RNA, particular for that at the 3'terminus; however, size and shape of the initiator appear to influence its effectiveness (Winters and Edmonds 1973b; Tsiapalis et al. 1973; Edmonds 1982). Some inhibitors of poly(A) polymerase are known which have proven to be useful tools in studying the biological significance of 3'poly(A) addition to mRNA, e.g., cordycepin 5'-triphosphate (competitive inhibition; Müller et al. 1977c), coformycin 5'-triphosphate (noncompetitive inhibition; Schröder et al. 1981) and 9-β-D-arabinofuranosyladenine 5'-triphosphate (competitive inhibition predominantly of nuclear matrix-associated poly(A) polymerase; Schröder et al. 1984c; Rose and Jacob 1978).

II. Poly(A) Catabolic Enzymes

Poly(A) degradation is catalyzed by the following three enzymes all of which have been detected and purified to homogeneity by our group (recently reviewed in: Müller et al. 1983): endoribonuclease IV (EC 3.1.26.6; Müller 1976), endoribonuclease V (EC 3.1.27.8; Schröder et al. 1980a), and poly(A) specific 2',3'-exoribonuclease (EC 3.1.13.4; Schröder et al. 1980b)(see Table 1).

1. Endoribonuclease IV

The first poly(A) specific nuclease, which we discovered already in 1975 (Müller et al. 1975), was termed endoribonuclease IV (EC 3.1.26.6). The enzyme was purified from chick oviduct to homogeneity (Müller 1976). None of the presently known endoribonucleases possess such a high specificity for poly(A) as endoribonuclease IV (see Sect. C).

a) Purification of the Enzyme

Endoribonuclease IV was isolated from chick oviducts and purified nearly to homogeneity (Müller 1976). By ammonium sulfate precipitation, DEAE-cellulose chromatography, phosphocellulose-chromatography, gelfiltration on Sephadex G-100, and hydroxyapatite chromatography, a 15,000-fold increase in the specific activity in a 25% yield could be achieved. The poly(A) anabolic poly(A) polymerase was found to copurify with the nuclease during chromatography on DEAE-cellulose and phosphocellulose, and could be removed from endoribonuclease IV only by chromatography on hydroxyapatite at a pH of 8.2.

b) Properties of the Enzyme

Endoribonucleae IV consists of a single polypeptide chain with a molecular weight of about 45,000. The enzyme specifically degrades poly(A) under formation of oligonucleotides of an average chain length of 10 nucleotide units which are terminated by 3'-hydroxyl and 5'-phosphate groups; one enzyme molecule causes 1.7×10^3 cleavages per min. The poly(A) segment of polysomal poly(A)(+)mRNA is selectively hydrolyzed by endoribonuclease IV up to a residual oligo(A) stretch of 5 AMP residues which is not degraded by the enzyme; the information-bearing stretch of the mRNA molecule remains unaffected. The purified enzyme exhibits optimal activity in Tris-HCl buffer at pH 8.7 and requires divalent cations (Mg^{2+} or Mn^{2+}) for activity. Mn^{2+} is the more effective cation. In the presence of EDTA the enzyme activity is completely inhibited. The optimal temperature is approx. 37°C. The endoribonuclease IV does not disintegrate the polyribosomal complex in vitro and has no influence on the translational capacity of a cell-free protein-synthesizing system. The enzyme seems to be localized predominantly in the nuclei (Müller et al. 1976, 1977a; Warnick and Lazarus 1977); however, a less enriched enzyme fraction from rat liver microsomes with similar properties has been described (Kumagai et al. 1980). In the quail oviduct, the enzyme is present in a concentration of about 4×10^4 molecules per cell.

c) Complex Formation with Poly(A) Polymerase

A peculiar property of endoribonuclease IV is its high affinity to poly(A) polymerase. The two enzyme molecules form a complex which can be separated by chromatography on hydroxyapatite at a pH of 8.2 (Tsiapalis et al. 1975; Müller 1976) and which remains stable at pH 6.8 (Abraham and Jacob 1978). The functional significance, if there is one, of the complex formation between the poly(A) anabolic and the poly(A) catabolic enzyme is not known. In nuclear envelopes from rat liver, which contain a considerable part of the intracellular endoribonuclease IV activity, no poly(A) polymerase activity could be detected (Bernd et al. 1982b) indicating that both enzymes can exist in some cellular compartments as separate entities.

2. Endoribonuclease V

Besides Endoribonuclease IV, a second poly(A)-degrading, endoribonuclease could be detected by us and purified to homogeneity from calf thymus which was called endoribonuclease V (EC 3.1.27.8; Schröder et al. 1980a). This enzyme selectively hydrolyzes poly(A) and also poly(U). In contrast to endoribonuclease IV, endoribonuclease V represents an endo-exonuclease producing 3'-AMP as end product.

a) Purification of the Enzyme

Endoribonuclease V was purified from calf thymus by acid precipitation, two successive chromatographic steps on ion exchange

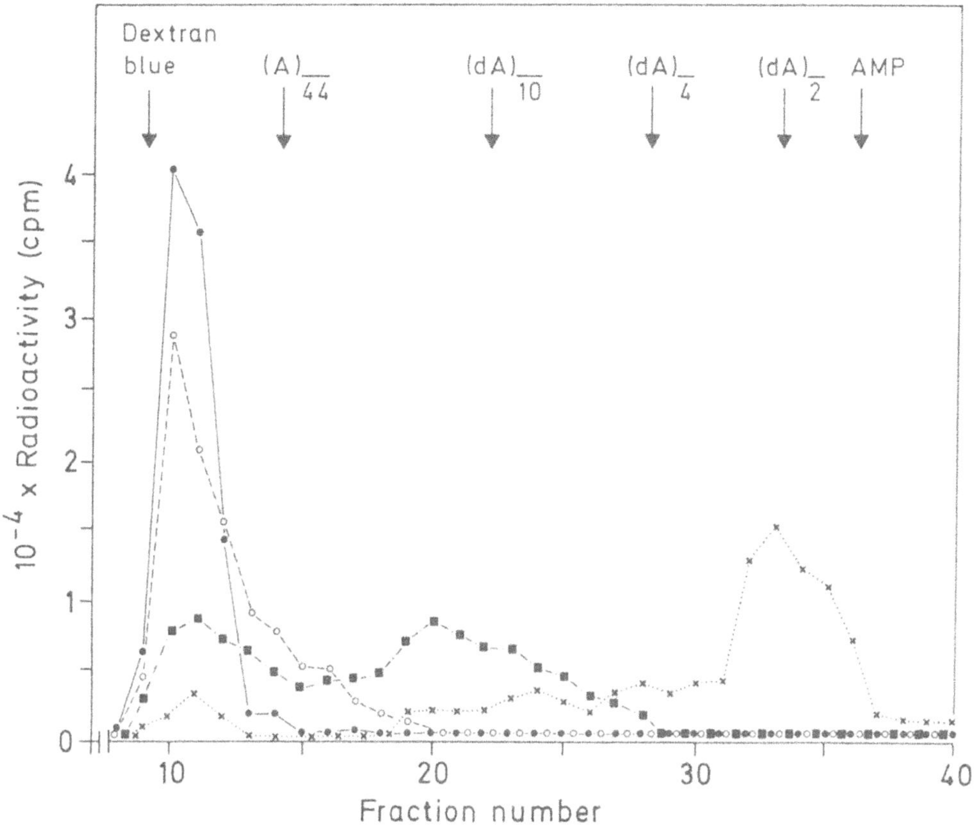

Fig. 2. Chain lengths distribution of the reaction products released from poly(A) by endoribonuclease V after different time intervals analyzed by gel filtration on Sephadex G-50. [^{3}H]Poly(A) was incubated with endoribonuclease V in the standard assay for 0 min (●), 10 min (o), 30 min (■) and 120 min (x). The *arrows* mark the positions of the authentic compounds. (Schröder et al. 1980a)

columns, gelfiltration, and affinity elution (Schröder et al. 1980a). During the final step, the enzyme was adsorbed onto a hydroxyapatite column, and eluted with 5'-AMP. The enzyme preparation was found to be homogeneous, based on analysis by polyacrylamide gel electrophoresis in the absence or presence of sodium dodecyl sulfate and isoelectric focusing.

b) Properties of the Enzyme

The endoribonuclease V cleaves preferentially single-stranded poly(A) and at a slower rate also poly(U) (Schröder et al. 1980a). None of the other single- as well as double-stranded synthetic polyribo- and polydeoxyribonucleotides tested were degraded; rRNA is hydrolyzed with an efficiency of 33% compared to poly(A) (= 100%). The pure enzyme degrades poly(A) to produce first oligoribonucleotides and ultimately 3'-AMP (Fig. 2); no formation of P-2':3'-Ado was detected. One molecule of endoribonuclease V causes 3.7×10^3 cleavages per min. The mol. wt. of the enzyme which seems to be composed of one polypeptide chain was determined to be 52,300. The enzyme has a pH optimum of 5.8, an isoelectric point of pH 6.3 and requires neither monovalent nor divalent cations. The temperature optimum is 37°C. The endoribonuclease V activity is strongly dependent on the ionic strength; addition of 150 mM NaCl abolished the enzyme activity by approximately 50%. The enzyme is insensitive to the thiol group reagent N-ethylmaleimide. The intracellular localization of endoribonuclease V is not known. The enzyme concentration has been estimated to be 1.6×10^5 molecules per cell.

3. Poly(A)-specific 2',3'-exoribonuclease

In 1977 we discovered a poly(A) degrading exoribonuclease in crude extracts from L5178y mouse lymphoma cells (Müller et al. 1977a) which shows similar properties to an enzyme described by Lazarus and Sporn already 1967 (Lazarus and Sporn 1967). This enzyme was later purified by us to homogeneity from calf thymus (EC 3.1.13.4; Schröder et al. 1980b); the purified enzyme has been proven specifically to cleave poly(A) under formation of 5'-AMP (3'-exonucleolytic activity). Later we found that the exoribonuclease degrades not only oligoadenylates with 3',5'-phosphodiester linkages but also those with unusual 2',5'-internucleotide bonds (Müller et al. 1980a). These 2',5'-linked oligoribonadenylates were found to be synthesized from ATP by a synthetase enzyme activated in interferon-treated cells (Kerr and Brown 1978). Therefore, this enzyme can be classified as a 2',3'-exoribonuclease.

a) Purification of the Enzyme

The poly(A)-specific 2',3'-exoribonuclease was isolated and purified to homogeneity from calf thymus gland by a procedure achieving an 8300-fold increase in specific activity at a 9.1% yield (Schröder et al. 1980b). The purification scheme includes the following steps: acid precipitation, DEAE-cellulose chromatography, preparative gel electrophoresis, ammonium sulfate precipitation, and affinity chromatography on poly(A)-Sepharose. Binding of the exoribonuclease to the affinity matrix during the final purification step was found to take place only in the presence of divalent cations in the binding buffer. However, in the presence of Mn^{2+} or Mg^{2+} the enzyme was found to destroy the poly(A) coupled to the matrix to a considerable extent. Therefore, we searched for a cation which was able to allow sufficient binding of the nuclease to the column without degradation of the ligand; such an ion was found to be Ba^{2+}. The enzyme preparation which was electrophoretically pure, could be stabilized in the presence of glycerol and D-galactose in the storage buffer.

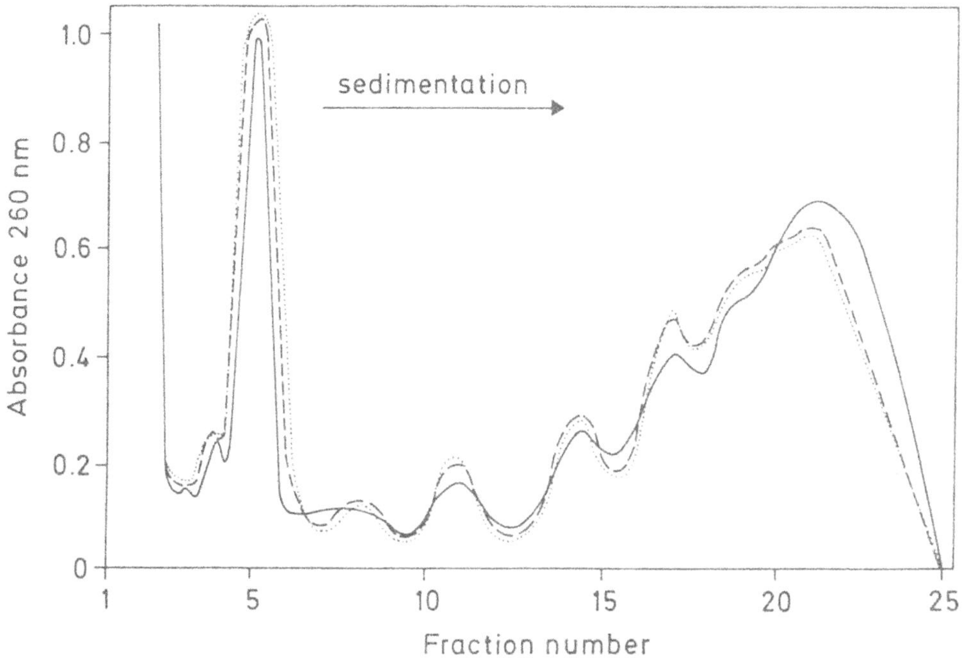

Fig. 3. Effect of poly(A)-specific 2',3'-exoribonuclease on polysome profile. Polysomes were isolated from quail oviduct (Müller et al. 1976). Polysome profile obtained after incubation in the absence (---) and in the presence (····) of exoribonuclease; polysome profile of the nonincubated sample (——). Due to an endogenous RNA-splitting activity a part of the large polyribosomes of the nonincubated preparation disintegrated after incubation into smaller entities. Note that there is no difference between the polysome profiles of the samples incubated with or without 2',3'-exoribonuclease. (Schröder et al. 1980b)

b) Properties of the Enzyme

The poly(A)-specific 2',3'-exoribonuclease has a mol. wt. of about 80,000 as estimated by gel filtration and consists of two subunits with mol. wts. of 58,000 and 31,000 as analyzed by sodium dodecyl sulfate-polyacrylamide gel electrophoresis (Schröder et al. 1980b). The enzyme has an absolute requirement for divalent cations; optimal enzyme activity is measured either in the presence of 3 mM $MgCl_2$ or 0.5 mM $MnCl_2$. Addition of 150 mM NaCl or 10 mM phosphate inhibited the enzyme activity by about 25%. Optimal activity was observed under alkaline conditions in a narrow range of pH values (8.7 to 9.4; ammediol buffer). The optimum temperature was found to be approx. 39°C. The enzyme displays sensitivity to reagents for thiol groups; in the presence of 5 mM N-ethylmaleimide or 7 mM o-iodobenzoate, the reaction was inhibited by approximately 80%. The exoribonuclease specifically cleaves poly(A) in the single- or in the double-stranded form [poly(A)·poly(U)]. The only product formed by the action of the enzyme is 5'-AMP. The K_m of the exoribonuclease for poly(A)

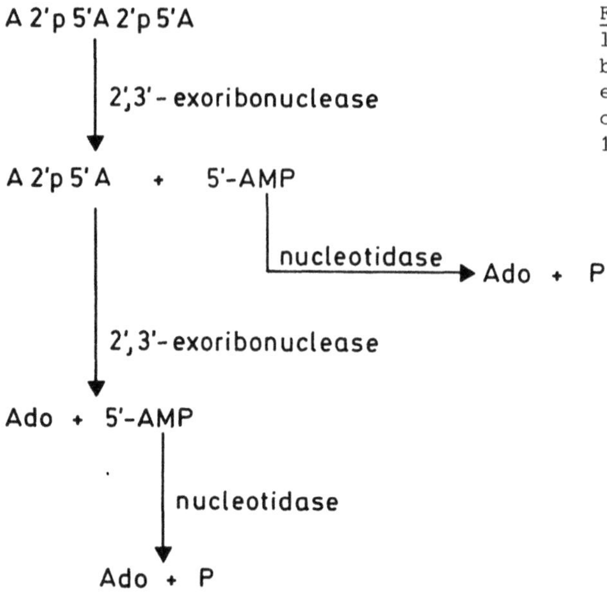

Fig. 4. Degradation of 2',5'-linked oligoriboadenylates by poly(A)-specific 2',3'-exoribonuclease and 5'-nucleotidase. (Müller et al. 1980a)

was determined to be 5.1 μM. The exoribonuclease does not disintegrate polyribosomes isolated from quail oviduct (Fig. 3); this finding is a strong indication that the exoribonuclease does not degrade the non-poly(A) stretches of mNRA's. In L5178y mouse lymphoma cells, the poly(A)-specific exoribonuclease was found of be predominantly localized in the nuclei (Müller et al. 1977a).

c) Involvement in 2',5'-oligo(A) Metabolism

Oligoriboadenylates with 2',5'-internucleotide linkages [ppp(A2'p)$_n$A (n = 2 to about 10)] were first found in interferon-treated cells and rabbit reticulocyte lysates (Hovanessian et al. 1977; Kerr and Brown 1978), and later also in isolated nuclei of untreated HeLa cells (Nilsen et al. 1982b). They are synthesized from ATP by a 2',5'-oligo(A) synthetase (Hovanessian et al. 1977) which is inducable by interferon (Ball 1979) and activated by double-stranded RNA (Kerr and Brown 1978). In the 5'-triphosphate (or 5'-diphosphate) stage, these oligoadenylate species are able to activate a latent cellular endonuclease (RNase L or F; Schmidt et al. 1978; Sen 1982) which hydrolyzes viral (Baglioni et al. 1979) and cellular (Sen 1982) single-stranded RNA primarily after AU, UG, and UU sequences (Floyd-Smith et al. 1981; Wreschner et al. 1981), resulting in inhibition of protein synthesis (Kerr and Brown 1978). The unphosphorylated 2',5'-oligo(A) "core" analogues, on the other side, were found to cause in synchronized cells inhibition of DNA synthesis (antimitogenic effect; Floyd-Smith et al. 1981). However, the biological activity of 2',5'-oligo(A) is limited by the rapid degradation of the oligomers (Minks et al. 1979; Williams et al. 1978). Cleavage of 2',5'-oligo(A) occurs from the 3'-terminus by a relatively unspecific 2'-phosphodiesterase (Schmidt et al. 1979) or by the adenylate-specific 2', 3'-exoribonuclease (Schröder et al. 1980b; Müller et al. 1980a). In Figure 4 the enzymatic degradation steps of the trinucleotide A2'p5'A2'p5'A by 2',3-exoribonuclease and 5'-nucleotidase (Müller et al. 1977b) are shown. In the first step A2'p5'A2'p5'A is cleaved to the dimer A2'p5'A and 5'-AMP. After a further step the dimer is hydrolyzed to Ado and 5'-AMP. In the presence of 5'-nucleotidase a complete degradation to Ado occurs.

Recently we found (Schröder et al. 1984b) that the xyloadenosine analogue of the 2',5'-oligo(A) core (xyloA2'p)$_2$xyloA (Gosselin and Imbach 1981), is not cleaved by 2',3'-exoribonuclease. Among the synthetic 2',5'-oligo(A) analogues that had been synthesized to improve its metabolic stability this compound has been proven most stable and active when tested in cell extracts (Eppstein et al. 1983).

III. Regulation of Poly(A) Metabolic Enzyme

The adjustment of the appropriate size of the poly(A) segment of hnRNA or mRNA actually required for its function, e.g., ensuring the metabolic stability of the RNA chain (Sheiness et al. 1975; Marbaix et al. 1977) or facilitation of nucleocytoplasmic translocation of mRNA (Schwartz and Darnell 1976; Bernd et al. 1982b) demands regulatory mechanisms controlling the activities of the poly(A) anabolic and poly(A) catabolic enzymes. These control processes are recognized in first outlines and they may be performed, firstly, by poly(A)-associated proteins (Blobel 1973; Müller et al. 1978a), secondly, by modified nucleotides in the poly(A) tract (Nichols and Welder 1981), thirdly, by oligo(U) sequences (Kish and Pederson 1977) which are detected in mRNA as well as hnRNA in a size of approximately 30 nucleotides (Molloy et al. 1972; Korwek et al. 1976; Wood and Edmonds 1981; Wallace et al. 1981), fourthly, by cytoskeletal proteins (Schröder et al. 1982b), and fifthly, by posttranslational modification of the poly(A) metabolizing enzymes (Rose and Jacob 1979, 1980; Tsiapalis et al. 1982; Schröder et al. 1983). In the following Sections, the control of polyadenylation and polydeadenylation of hnRNA and mRNA by posttranslational enzyme modification, poly(A)-associated proteins, and tubulin and G-actin will be described in more detail.

1. Posttranslational Modification

Poly(A) chain length regulation could be performed by posttranslational protein phosphorylation/dephosphorylation processes of the poly(A) metabolic enzymes (Rose and Jacob 1979, 1980; Tsiapalis et al. 1982). Some disagreement exists about which component of the poly(A) polymerase-endoribonuclease IV complex is phosphorylated. Rose and Jacob (1979, 1980) have reported poly(A) polymerase from liver and tumor cell nuclei to be phosphorylated by a specific nuclear kinase, while Tsiapalis et al. (1982) showed that in the poly(A) polymerase-endoribonuclease IV complex only the poly(A) degrading enzyme could be phosphorylated by a soluble protein kinase. The difference between the results of Jacob and Tsiapalis could be caused by the fact that both groups did not work with homogeneous enzyme preparations. Using the homogeneous enzymes we could recently demonstrate (Schröder et al. 1983) that both poly(A) anabolic poly(A) polymerase and poly(A) catabolic endoribonuclease IV are phosphoproteins which could be posttranslationally controlled by phosphorylation by nuclear protein kinase NI (Baydoun et al. 1982). Both enzymes are not phosphorylated by cytoplasmic, microtubuli-associated cAMP-dependent protein kinase from calf brain. Phosphorylation of poly(A) polymerase (Rose and Jacob 1980) or endoribonuclease IV (Tsiapalis et al. 1982) has been proposed to result in a higher specific activity of the respective enzymes. Recently we could present evidence (Schröder et al. 1983) that the decreased amounts of short-sized oligo(A) fragment during S phase of synchronized L5178y mouse lymphoma cells and during ageing (quail oviduct) could in part be due to posttranslational modification of enzymes involved in poly(A) metabolism.

2. Poly(A)-associated Proteins

Specific poly(A)-associated proteins have been shown to interact with the poly(A) segment of mRNA with formation of the poly(A)-ribonucleoprotein [poly(A)-RNP] complex (Kwan and Brawerman 1972; Jeffery 1978; Vincent et al. 1981; Bernd et al. 1982c). Analysis of the poly(A)-RNP complex revealed that it might consist of one to six polypeptide species with an M_r = 75,000 protein as the major component (Blobel 1973; Vincent et al. 1981; Jeffery 1977; Tomcsányi et al. 1981; Adams et al. 1980; Müller et al. 1978a; Bernd et al. 1982a,c). By selective radiolabeling of the protein components in the polysomal poly(A)-RNP from L5178y mouse lymphoma cells with [^{3}H]dansyl chloride, two protein species with M_r = 77,000 (P77) and M_r = 54,000 (P54) have been identified (Bernd et al. 1982c). A quantitative analysis revealed that the poly(A)-RNP complex is composed of 3.7 molecules P54 and 1.9 molecules P77. Digestion experiments with dimers and tetramers containing ribonuclease A indicated that P54 covers 15 - 20 AMP residues and P77 a sequence of 40 - 45 nucleotides on the poly(A) 155 stretch of poly(A)-RNP. Using different approaches, Adams et al. (1980) reported that a 45 nucleotide tract is covered by the poly(A)-associated proteins, while Baer and Kornberg (1980) determined fragments of 22 AMP units. In some reports the presence of P73-77 in hnRNP has been described (Kish and Pederson 1975), while in others (Tomcsányi et al. 1983) it was shown that P63 is the major poly(A)-associated protein in the nucleus. Van Eekelen et al. (1981) demonstrated that P73 was associated with poly(A) in HeLa cell cytoplasm but not in the nuclei.

It was suggested (Rose et al. 1979) that poly(A) polymerase from Morris hepatoma 3924 A is antigenically related to the poly(A)-associated protein P75. The authors assumed that the poly(A) polymerase protein (M_r = 60,000) and the P75 protein (M_r = 75,000) have similar or perhaps identical antigenic determinants. However, analytical data presented by us (Müller et al. 1983) do not support these immunological data.

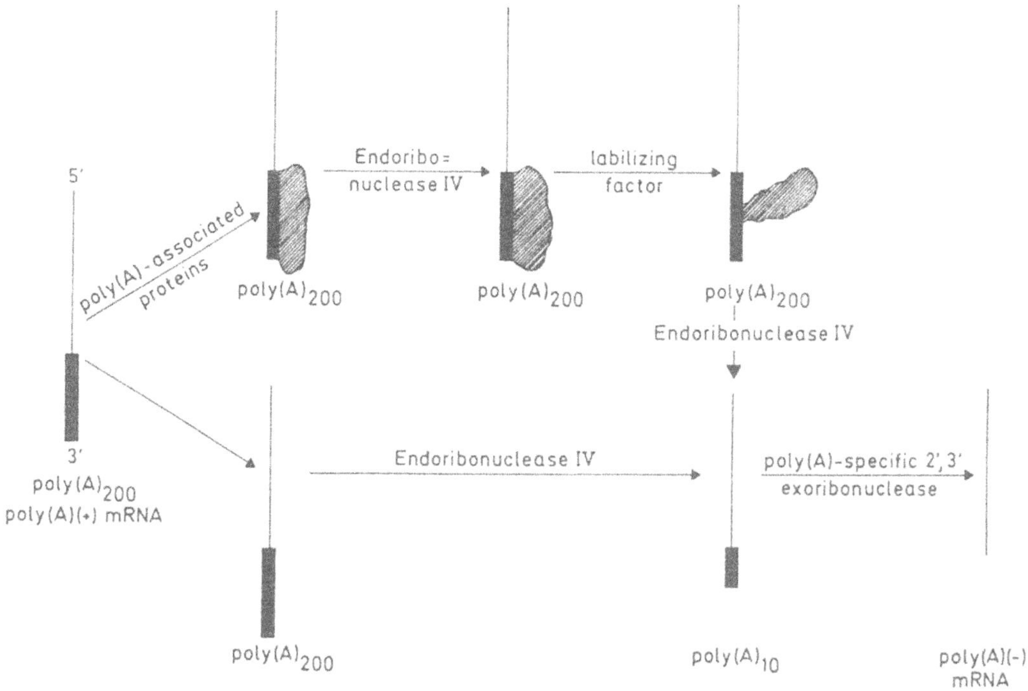

Fig. 5. Control of ribonuclease-mediated poly(A) degradation by poly(A)-associated proteins. (Müller et al. 1980b)

In a recent report, we could establish a correlation between the loss of poly(A)-associated proteins and the size of the poly(A) tail of mRNA (Bernd et al. 1982a). Using the quail oviduct model, the number of poly(A)-associated proteins per poly(A) tail was determined to be reduced from 4.7 to 1.9 molecules during ageing. Concomitantly with this loss, a reduction of the size of the poly(A) segment of mRNA occurs from 120 - 180 AMP units to 70 AMP units.

It has been assumed that P77 is involved in nucleocytoplasmic mRNA transport (Schwartz and Darnell 1976; Roy et al. 1981). Moreover, evidence has been presented recently that the poly(A)-associated protein P76 is involved in the formation of the 48S initiation complex during protein synthesis (Schmid et al. 1983). In an enzymic study, we determined the protection functions of P77 and P54 for the poly(A) segment of both polysomal poly(A)(+) mRNP and isolated poly(A)-RNP complex towards an attack of endoribonuclease IV and of the poly(A)-specific 2',3'-exoribonuclease (Müller et al. 1978a). It was found that the exoribonuclease hydrolyzes the poly(A) segment in poly(A)(+)mRNP and poly(A)-RNP irrespectively of the presence of poly(A)-associated proteins. In contrast, these proteins protect the poly(A) segment almost completely from the nucleolytic attack of the endoribonuclease IV; after removal of these proteins by 1 M urea the poly(A) segment is highly susceptible to degradation by endoribonuclease IV.

Thus we propose that degradation of poly(A) occurs in two steps: firstly, endonucleolytic digestion by endoribonuclease IV, controlled by the poly(A)-associated proteins, and secondly, exonucleolytic hydrolysis of the residual 3'-terminal oligo(A)$_{10}$ stretch not degraded by endoribonuclease IV (Müller et al. 1976) and of the released oligo(A)$_{10}$ fragments by the poly(A)-specific 2',3'-exoribonuclease regulated by a hitherto unknown mechanism (Müller et al. 1980b); see Figure 5. In mammalian cells, the existence of a cytoplasmic "labilizing factor" has been reported by Bergmann and Brawerman (1977), which interferes with the poly(A)-poly(A)-associated protein interaction, thus rendering the poly(A) chain susceptible to degradation by endoribonuclease IV (Fig. 5).

3. Interaction with Cytoskeletal Elements

It has been shown that RNA binds to microtubule protein in vitro (Bryan et al. 1975; Schröder et al. 1982a, 1984a) and hnRNA is attached to proteins of the nuclear matrix (Van Eekelen and van Venrooij 1981). Therefore, we studied the effect of microtubule protein and actin on the activities of the poly(A) metabolizing enzymes (Schröder et al. 1982b).

The activities of poly(A) polymerase and of endoribonuclease IV from calf thymus are strongly inhibited after preincubation with tubulin dimers, polymerized tubulin, and G-actin. Concerning the poly(A) polymerase-endoribonuclease IV complex, the highest inhibitory effect was measured after preincubation with tubulin dimers (64% reduction of the activity of poly(A) polymerase and 47% reduction of endoribonuclease IV activity); somewhat lower was the inhibitory potency of polymerized tubulin and G-actin. On molar basis, 50% enzmye inhibition was observed at a ratio of 0.7:1 (tubulin dimer/enzyme) and 1.9:1 (G-actin/enzyme). The complex formed between the two poly(A)-specific enzymes and the contractile proteins was found to possess a stoichiometry of 1:1 as determined by gel chromatography (Schröder et al. 1982b).

The poly(A) polymerase, separated from the complex and purified to homogeneity, showed the same sensitivity toward the contractile proteins as the enzyme associated with endoribonuclease IV. As in the complex form, the homogeneous enzyme was most sensitively inhibited by tubulin dimer (by 79%), followed by polymerized tubulin (72%) and G-actin (49%).

IV. Dependence on Physiological Factors

It is well established that the activities of the poly(A) metabolizing enzymes respond to altered physiological conditions and neoplasia. Some data on the poly(A) polymerase activity have been summarized by Jacob and Rose (1978). However, several lines of evidence indicate that primarily the activities of the poly(A) catabolic poly(A) nucleases, and not that of the poly(A) anabolic poly(A) polymerase, are correlated with the physiological state of the cell or the developmental stage of the organism. In the

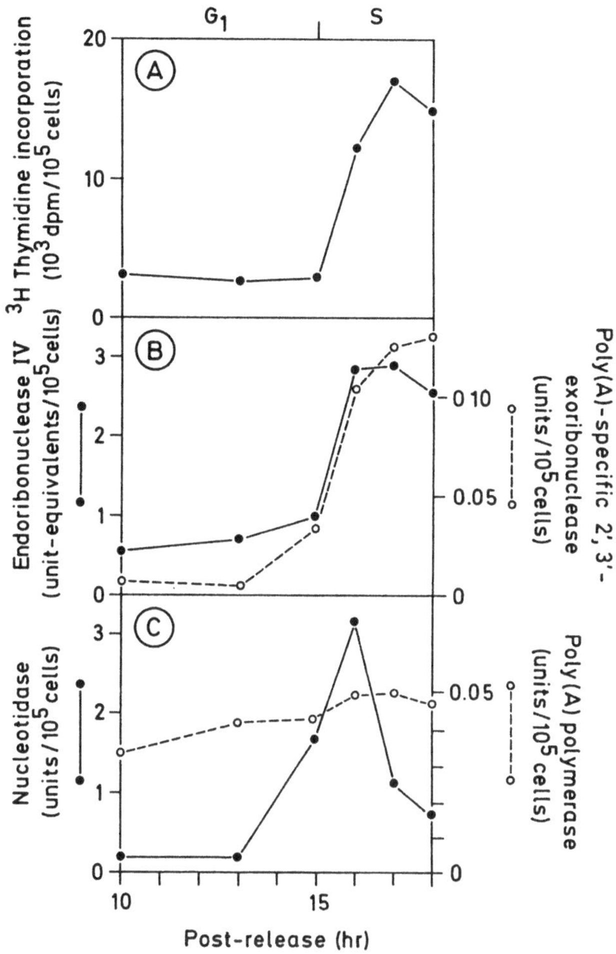

Fig. 6. Changes of extractable activities of poly(A) polymerase, endoribonuclease IV, poly(A)-specific 2',3'-exoribonuclease and 5'-nucleotidase during G_1 and S phase of synchronized L5178y mouse lymphoma cells. The rate of DNA synthesis was determined by [^{3}H]thymidine incorporation. (Müller et al. 1977a)

following chapters the influences of the stage of cell cycle, ageing, viral infection, and nutrition on poly(A) metabolizing enzyme activities will be discussed.

1. Cell Cycle

The poly(A) anabolic enzyme poly(A) polymerase as well as the poly(A) catabolic enzymes endoribonuclease IV and 2',3'-exoribonuclease has been shown to be primarily present in the nucleus of L5178y mouse lymphoma cells (Müller et al. 1977a). In addition, a nuclear 5'-nucleotidase which is highly specific for 5'-AMP (Müller et al. 1977b) has been reported to be involved in poly(A) catabolism by conversion of the product of the exoribonuclease reaction to adenosine (Müller et al. 1977a). In an enzymic study, the alterations of the levels of the poly(A) metabolizing enzymes in synchronized L cells were determined (Müller et al. 1977a). It was found that the extractable activities of the poly(A) catabolic enzymes markedly increase parallel with the increase of

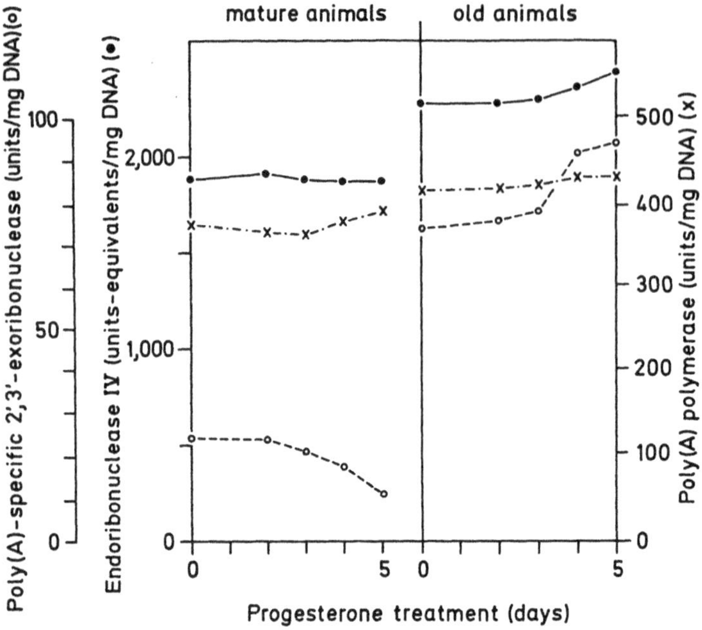

Fig. 7. Age-dependence and hormone-dependence of the levels of poly(A) polymerase and poly(A) nuclease in quail oviduct. Mature (postnatal age of 70 days) and senescent (950 days) animals were injected for 0 - 5 days with progesterone (50 mg/kg body wt.). Subsequently, the oviducts were removed, extracted, and the enzyme activities were determined. (Müller et al. 1979)

DNA synthesis; the endoribonuclease IV activity increases fourfold during the transition from G_1 to S phase, the poly(A)-specific 2',3'-exoribonuclease and the nucleotidase activity 30-fold and 16-fold respectively (Fig. 6). On the other hand, the activity of poly(A) polymerase remains essentially constant during G_1 to S phase. This means that during S phase a change of the balance between poly(A) anabolic and poly(A) catabolic enzyme activities occurs in favor of the poly(A) degradation. Interestingly, most histone mRNA's, which are synthesized during S phase, lack poly(A) (Adesnik and Darnell 1972; Stein et al. 1975). However, the possibility that histone mRNA's are initially polyadenylated, but subsequently undergo a rapid deadenylation cannot be excluded. Although lacking the nonhistone AAUAAA poly(A) initiation site (Proudfoot and Brownlee 1976), some histone mRNA's are found to be polyadenylated (Levenson and Marcu 1976; Ruderman and Pardue 1977). The high level of 2',3'-exoribonuclease in S-phase cells may also explain the recently obtained analytical result (Schröder et al. 1983) that the relative amount of oligo(A)$_{2-6}$ sequences during S-phase of synchronized L-cells is five times lower than that in the stationary L-cell system.

2. Ageing

Studying the age-correlated alterations of poly(A) polymerase and poly(A) nucleases in quail oviduct, it was found (Müller et al. 1979) that the levels of extractable poly(A) polymerase and endoribonuclease IV are only a little higher in oviducts from senescent animals compared to adult animals (Fig. 7). Furthermore, the activities of both enzymes do not change significantly after progesterone stimulation (Müller et al. 1975) of quails of both age groups. The activity of the poly(A)-specific 2',3'-exoribonuclease, however, increases markedly during ageing: the activity of this enzyme raised 3.5-fold from 22 units per mg DNA tissue in mature quails to 75 units per mg DNA tissue in senescent animals (Fig. 7). Moreover, after a 5-day progesterone treatment the exoribonuclease activity decreases slightly in mature animals, while it increases in senescent ones by 30%. The conclusion drawn from these data that the poly(A) segment of mRNA is shorter in senescent animals compared to adult ones, is supported by analytical results. Determinations of the average sizes of the poly(A) sequences of mRNA's from oviducts of mature and old quails revealed (Bernd et al. 1982a) that the length of the poly(A) segment is shortened from an average value of 130 AMP units (adult animal group) to an average value of 70 AMP units (senescent animal group). Concomitantly a drop of poly(A)-associated protein molecules from approximately 4.7 molecules (mature oviduct) to 1.9 molecules (old oviduct) could be detected (Bernd et al. 1982a). Recently we could show (Schröder et al. 1983) that the relative amounts of low mol. wt. oligo(A) fragments formed in vivo in oviducts of mature and old animals gradually decrease with ageing of the animals; the percentage of oligo$(A)_{2-6}$ sequences in the mature animal group was found to be six times higher than that in the old animal group. This result can be explained by the fact that due to the higher content of poly(A) exoribonuclease, oligo(A) fragments, once formed, are rapidly degraded in organs of older animals. The impaired post-transcriptional net polyadenylation of hnRNA during ageing due to an elevated level of the poly(A) catabolic 2',3'-exoribonuclease at an essentially unchanged poly(A) anabolism may result in reduced mRNA stability and reduced nucleocytoplasmic mRNA transport and, perhaps, in a reduced splicing efficiency. Recently, Khasigov et al. (1983) demonstrated that the half-life time of hnRNA is increased with ageing possibly due to a deceleration of hnRNA maturation.

3. Nutrition

The balance between poly(A) anabolic and poly(A) catabolic enzymes seems to be also nutritionally regulated. Studying the extractable enzyme activities from liver nuclei of fed and fasted rats, Matts and Siegel (1979) reported a three- to fourfold increase in the level of poly(A) endonuclease in extracts from fasted animals when compared to fed rats; on the other hand, no differences in the levels of poly(A) polymerase and poly(A) exoribonuclease activity were found in the two animal groups. The fasting induced increase in the poly(A) endonuclease activity could be reversed by refeeding certain amino acids (alanine, asparagine, glutamine, isoleucine, proline, and serine) or blocked by adrenalectomy. In the presence of corticosterone, however, the fasting-induced effect was also obtained in adrenalectomized rats. It is interesting to note that feeding also causes a reduction of the nuclear envelope NTPase-mediated nucleocytoplasmic transport of poly(A)(+)mRNA (Murty et al. 1977) which can be reversed by L-tryptophane (Murty et al. 1982).

4. Virus Infection

The activities of poly(A) metabolic enzymes are also altered in cells after virus infection. The extractable activities of the poly(A) catabolic enzymes in BHK cells, infected with herpes simplex virus (type 1; strain Lennette), were found to increase already after 1 - 2 h p.i. reaching a maximum at 4 - 9 h p.i. (Müller et al. 1978b); the maximal increase in the activity of endoribonuclease IV is fourfold and that of the poly(A)-specific 2',3'-exoribonuclease 15-fold. Unlike the time course of the poly(A) nuclease activities which coincides with the appearance of viral DNA, the onset of the increase of poly(A) polymerase activity occurs later; maximal enzyme activity (16-fold increase) is obtained 9 - 10 h p.i. (Müller et al. 1978b,c).

E. Splicing

Exon-intron and intron-exon boundaries within the hnRNA are characterized by highly conserved sequences. These consensus sequences have been identified as (C,A)AG GU(A,G)AGU for the 5'splice site and (U,C)$_n$N(C,U)AG/G for the 3'splice site (sequences belonging to the exons are underlined).

Comparison of the consensus sequences at RNA splicing sites in hnRNA with sequences in U1 RNA revealed an exact complementarity (Lerner et al. 1980; Rogers and Wall 1980), suggesting a model in which U1 RNP is one recognition component of the nuclear RNA splicing apparatus; in this model, U1 RNA helps to juxtapose the two splice junctions. This assumption is supported by the finding that snRNP's are directly associated with hnRNA (Gallinaro et al. 1981; Prüsse et al. 1982). It has been demonstrated that antiserum from patients with systemic lupus erythematosus, which precipitates the U1 RNP and the Sm antigen, inhibits splicing of adenovirus mRNA precursors in isolated HeLa cell nuclei (Yang et al. 1981). These results were recently confirmed by Padgett et al. (1983) using a soluble in vitro system for adenovirus mRNA splicing. While the specificity for the splicing reaction is partially encoded in the consensus sequences at the 5' and 3' splice sites (Sharp 1981; Mount 1982), the ones for splicing of rRNA precursors (Kruger et al. 1982) and pre-tRNA (Peebles et al. 1983) are provided by the secondary and tertiary structures of these RNA species.

I. SnRNP's and scRNP's

SnRNA's have been found in all vertebrate tissues so far investigated (for reviews see Busch et al. 1982; R. Reddy and Busch 1983; Brunel et al. 1984). Intracellularly they are found complexed with proteins as snRNP particles. The U snRNA's which are capped are synthesized just as the hnRNA's by RNA polymerase II, while the La and Ro RNA's are like tRNA and 5S RNA in that all are RNA polymerase III transcripts. U snRNP's containing U1, U2, U4, U5, and U6 snRNA are located in the nucleoplasm, whereas U3 snRNA resides in the nucleolus (Busch et al. 1982). The nucleoplasmic snRNP's are found at least in part to be associated with hnRNP particles (Sekeris and Niessing 1975). Their function may be in processes occurring during hnRNA splicing to mRNA (Lerner et al. 1980; Rogers and Wall 1980). Thus, the 5'terminal 20 base pair sequence of U1 RNA has been shown to display complementarity to the donor and acceptor splice site consensus sequences in hnRNA (Lerner et al. 1980). Therefore, it was suggested that U1 hnRNP's may align the splice sites during the hnRNA splicing process (Lerner et al. 1980; Rogers and Wall 1980). Yang et al. (1981) demonstrated splicing of adenovirus mRNA to be inhibited in nuclei incubated with anti-RNP antibodies. More recently, Mount et al. (1983) showed that U1 snRNA is selectively bound to the 5' splice site in vitro. On the other hand, nucleolar U3 snRNP, which is hydrogen bonded to rRNA precursors, seems to be involved in rRNA processing (Zieve and Penman 1976; Busch et al. 1982). The physiological function of U2, U4, U5, and U6 snRNP's are unknown, though U4 snRNP has been proposed to be implicated in cleavage of the hnRNA 3'end (Berget 1984).

By antibodies produced by patients with systemic lupus erythematosus two further classes of small ribonucleoproteins could be identified: La snRNP's and Ro scRNP's (Lerner et al. 1981). Anti-La sera have been shown to recognize also the RNP's of further products of RNA polymerase III as pre-5S RNA and pre-tRNA (Hendrick et al. 1981; Rinke and Steitz 1982). Thus, La protein has been proposed to be involved in transcription, processing, or nucleocytoplasmic transport of RNA polymerase III transcripts (Rinke and Steitz 1982). One La snRNP species, the 4.5S RNA, which displays homology to the highly repetitive Alu family DNA from mammalian cells, was found to be associated with nuclear poly(A)(+)hnRNA and also cytoplasmic poly(A)(+)mRNA, possibly by base-pairing between certain purine-rich sequences within the 4.5S RNA and the noncoding, pyrimidine-rich consensus sequences at the mRNA 3'terminus (Harada and Kato 1980).

In contrast to U RNP's and La RNP's which are localized predominantly in the cell nucleus, Ro RNP's are thought to be cytoplasmic RNP complexes (Lerner and Steitz 1981). The Ro scRNP's, which possess 5'-triphosphate termini, are less abundant than the U snRNP's and not so highly conserved (Hendrick et al. 1981). Mouse cells contain two (mY1 and mY2) and HeLa cells even five Ro RNA species (hY1 - hY5) (Lerner et al. 1981; Hendrick et al. 1981). Ro scRNP's are suggested to be involved in mRNA translation or transport (Wolin and Steitz 1983) or degradation of spent mRNA's in the cytoplasm (Lerner and Steitz 1981).

II. Nucleases Possibly Involved in Splicing

Despite the efforts that have been undertaken, only little is known about the ribonucleases involved in the cleavage-ligation process of mRNA splicing in eukaryotes (Cech 1983). Only some nucleases from prokaryotes responsible for the maturation of tRNA and rRNA precursor molecules have been studied in some detail (see Sect. C). However, the demands made for a nuclease involved in mRNA splicing may be not the same as for tRNA or rRNA processing enzymes, e.g., in the latter cases the recognition of the secondary or tertiary structure of the RNA precursor seems to be important, while the specificity of the mRNA splicing reaction may base primarily on the recognition of the primary structure of the 3' and 5' donor and acceptor splice sites by U1 snRNP's.

In 45S particles we detected a novel pyrimidine-specific endoribonuclease, which was called endoribonuclease VII (Bachmann et al. 1983b). This enzyme has been shown to possess the unique property that its activity is modulated by poly(A). Moreover, besides the basic endoribonuclease VII (pI 8.5) 45S particles have been found to contain an acidic (pI 4.1) endoribonuclease which specifically degrades poly(U) (Bachmann et al. 1984a). Another distinct poly(U)-specific endoribonuclease has been identified in 12S U1 snRNP particles (Bachmann et al. 1983a). Although the physiological functions of endoribonuclease VII, pI 4.1 endoribonuclease and poly(U) specific U1 snRNP-associated endoribonuclease are not yet known, from their presence in RNP particles and their characteristics it is likely that these enzymes could play a functional role in RNA processing. Therefore, the properties of these enzymes will be described in the following Sections in more detail.

1. Endoribonuclease VII

In 1983 we reported the purification and characterization of a novel pyrimidine-specific 3'-endoribonuclease from calf thymus that fulfilled the requirements postulated for a splicing enzyme in a hypothesized poly(A)-mediated splicing process (Bachmann et al. 1983b). This enzyme which was termed endoribonuclease VII was found to be modulated by poly(A) and to be able to recognize oligo(U) sequences.

a) Purification of the Enzyme

Endoribonuclease VII was purified from calf thymus (Bachmann et al. 1983b) by ammonium sulfate precipitation of the crude extract followed by gel filtration on Sephadex G-150. During this procedure the high mol. wt. endoribonuclease VII fraction, which eluted between V_e/V_o values of 1.0 - 1.3, was separated from endoribonuclease IV (V_e/V_o, 1.73), endoribonuclease V (V_e/V_o, 1.65), poly(A)-specific 2',3'-exoribonuclease (V_e/V_o, 1.42) and ribonuclease A (pancreatic, EC 3.1.27.5; V_e/V_o, 2.36). After ion exchange chromatography on DEAE-cellulose and adsorption to phosphocellulose, the nuclease fraction contained two poly(U)-degrading enzyme activities which were distinguishable by their susceptibility to EDTA (degradation of poly(U) up to 3'-UMP in the absence of EDTA and in the presence of $MgCl_2$, and to 5'-phosphate terminated oligo(U) fragments in the presence of EDTA). It was subjected to preparative isoelectric focusing in a sucrose density gradient containing carrier ampholytes (pH 3.5 - 10). During this procedure a separation of both nuclease activities could be obtained; one peak at pH 8.5 (endoribonuclease VII) and a second

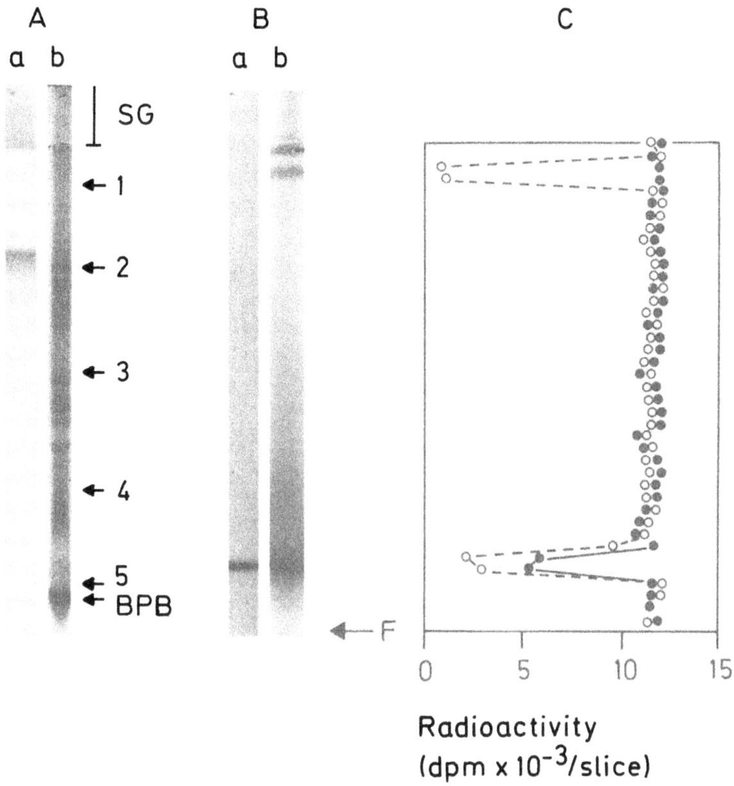

Fig. 8. Gel electrophoresis of endoribonuclease VII after centrifugation through a sucrose gradient. A mol. wt. estimation of proteins present in the endoribonuclease VII preparation on sodium dodecyl sulfate-polyacrylamide (10%) gels. Proteins of known mol. wt.: *1* phosphorylase a (M_r = 98,000); *2* bovine serum albumin (M_r = 68,000); *3* egg white albumin (M_r = 43,000); *4* chymotrypsinogen (M_r = 25,000); *5* cytochrome c (M_r = 12,500). *SG* stacking gel; *BPB* bromophenol blue. Fraction VI of the enzyme was used, which had been additionally fractionated by sucrose gradient centrifugation into top and bottom fractions. *Lane a* top fraction; *lane b* bottom fraction. B analysis of the enzyme preparations using 6 M urea-polyacrylamide (10% gels). *Lane a* top fraction obtained after sucrose gradient centrifugation; *lane b* bottom fraction. *F* front; C in situ detection of enzyme activity. Nuclease activity is shown in the separated top (*lane B, a;* •) and bottom fractions (*lane B, b;* o). (Bachmann et al. 1983b)

one at pH 4.1 (pI 4.1 endoribonuclease; see Sect. E.II.3). A calculation on the basis of the poly(U)-degrading activity (measurable in the characteristic pI range, pH 8.2 - 9.2, of the enzyme) revealed a 6340-fold purification of endoribonuclease VII.

b) Properties of the Enzyme

Based on analysis by velocity sucrose gradient centrifugation and gel electrophoresis under native and denaturing conditions

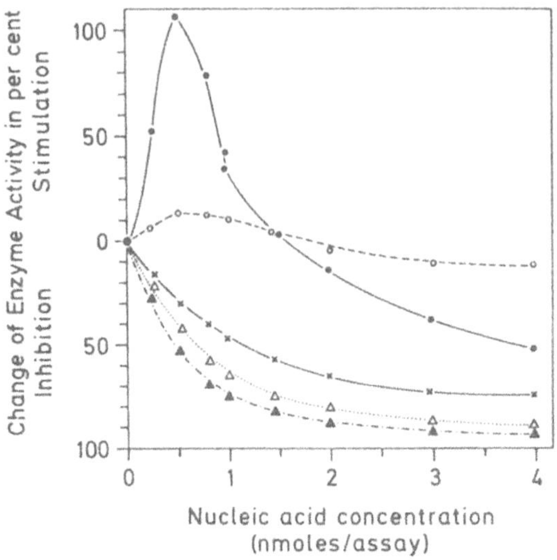

Fig. 9. Effects of poly(A), poly(G), and poly(dA) on endoribonuclease VII activity. The activity of purified endoribonuclease VII (Fraction VI) was determined in the presence of increasing amounts of poly$(A)_{95}$ (●), oligo$(A)_4$ (o), poly$(dA)_{120}$ (Δ), and poly$(G)_{85}$ (▲); the concentrations are given in nanomoles (with respect to phosphate)/assay. The percentage of stimulation or inhibition of the enzyme activities is indicated. For comparison, ribonuclease A activity in the presence of increasing concentrations of poly$(A)_{95}$ (x). (Bachmann et al. 1983b)

(see Fig. 8), the endoribonuclease VII was found to bind to nuclear 45S particles composed of RNA and 10 major proteins; one of these proteins (M_r = 74,000) is the enzyme itself. The enzyme has a pH optimum of 7.2 (Tris-acetate buffer), a temperature optimum of 37°C, an isoelectric point at pH 8.5 and requires neither monovalent nor divalent cations. Optimal enzyme activity is observed in the presence of 0.3 to 0.5 M urea. The enzyme cleaves poly(U) and poly(C) while other single-stranded homopolyribo- as well as polydeoxyribonucleotides are not degraded; poly(A,C) is hydrolyzed to a smaller extent, while poly(U)•poly(A) is not degraded at all. The K_m value for poly(U) as a substrate was found to be 11.3 µM. During hydrolysis, oligo$(U)_{11}$ fragments with 3'-OH and 5'-P termini are formed; this indicates that the enzyme is a 3'-endoribonuclease. One molecule of endoribonuclease VII causes 1.8×10^2 cleavages per min. The most prominent property of the enzyme is the fact that its poly(U)-degrading activity could be modulated by poly(A); see Figure 9. Poly$(A)_{95}$ stimulates the hydrolysis of poly(U) by the enzyme by more than 100% within a limited concentration range; at a molar ratio (based on phosphate content) of approx. 1 [poly(A)] : 10 [poly(U)], maximal stimulation of the enzyme activity was achieved; at lower ratios, 1:2 or 1:1, the hydrolysis of poly(U) was inhibited by 52% or 87%, respectively. Oligo$(A)_4$ fragments were found to have little modulating effect on the enzyme activity. On the other hand, poly(G) and poly(dA) strongly inhibit endoribonuclease VII. Binding studies revealed that endoribonuclease VII has a marked affinity for poly(A) and poly(U); no other homo- or heteropolymeric nucleic acids were detected to bind to the enzyme. Recently, an RNA fragment that was found to be tightly bound to the enzyme was identified to be oligo(U) (unpublished results). We could calculate that about 8% of the total protein of the 45S particle is endoribonuclease VII. Based on this figure the concentration of endoribonuclease VII can be estimated as 2.2×10^4 molecules per cell (Bachmann et al. 1983b).

c) A Hypothetical Model for Endoribonuclease VII Function

The physiological function of the poly(A)-modulated and oligo(U)-recognizing endoribonuclease VII is not yet known. The enzyme seems to be a constituent of an RNP complex that shows some homology to a snRNP particle based on its protein composition (Busch et al. 1982); thus, snRNP's are found to consist of 7 to 11 polypeptides with mol. wts. between M_r = 10,000 and M_r = 70,000 (Brunel et al. 1984) giving a protein pattern very similar to that found for the endoribonuclease VII complex. However, the sedimentation coefficients of both complexes are completely different; thus, the endoribonuclease VII complex has a sedimentation coefficient of 45S, while snRNP's sediment with 11 to 12 S (Brunel et al. 1981). At present we assume that a hypercomplex exists (Bachmann et al. 1983b). This seems to be likely because of the high affinity of snRNP's to 30S nuclear particles (Sekeris and Niessing 1985; Seifert et al. 1979) and to hnRNP monomers (Brunel et al. 1981). Moreover, 30 - 60S RNP complexes from mouse erythroleukemia cell nuclei have been shown to be composed of snRNP and hnRNP (Howard 1978). Based on its binding properties we assume that endoribonuclease VII is able to differentiate between oligo(U)(+)RNA's and oligo(U)(-)RNA's. HnRNA (Molloy et al. 1974) as well as mRNA (Wood and Edmonds 1981) have been shown to contain oligo(U) tracts within their molecules. Kish and Pederson (1977) proposed an in vivo association of the internal oligo(U) sequences with the 3'terminal poly(A) tract promoted by a binding protein. Thus, one possible function of endoribonuclease VII may be the recognition and the excision of the pyrimidine-rich sequences within the intron loops of hnRNA precursors during hnRNA processing. In an hypothetical model (Müller et al. 1984) the endoribonuclease VII-protein complex binds to the hnRNP at its 3'poly(A) sequence. This sequence has been proposed to align the splicing sites in hnRNA precursor via formation of a transitional conformation of the hnRNP particle stabilized by poly(A) triple strand formation with internal oligo(U) sequences (Bina et al. 1980). This assumption is supported by the finding that, under certain conditions, oligo(U) sequences in hnRNP polyparticles are complexed with poly(A). Moreover, the RNA present in the 45S particle have been shown to contain oligo(U) sequences (Bachmann, unpublished). The endoribonuclease VII is thought to recognize and subsequently excise the intron-loops formed. The adjacent free exon termini thereby created are sealed by an unknown RNA ligase perhaps also present in the hnRNP-snRNP hypercomplex. Another possible function for the enzyme may be the cleavage of the 3' terminal oligo(U) trailers common to RNA polymerase III transcripts, such as pre-tRNA (Melton et al. 1980), pre-5S RNA (Rinke and Steitz 1982) and Ro RNA precursors (Wolin and Steitz 1983). Analytical results revealed that during maturation of these RNA precursors their 3' oligo(U) sequences, consisting of 2 to 6 UMP residues, are progressively removed, most likely by an endoribonucleolytic mechanism (Melton et al. 1980; Rinke and Steitz 1982). Recently, Stefano (1984) demonstrated that the 3' terminal oligo(U) sequences of all RNA polymerase III transcripts represent a common binding site for the La protein.

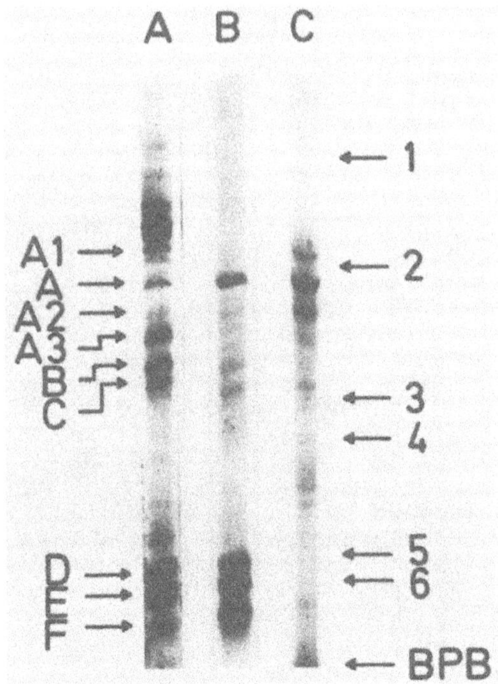

Fig. 10. Gel electrophoresis of the 12S particle containing the pI 4.1 endoribonuclease and of U snRNP. Electrophoresis was performed in sodium dodecyl sulfate - 10% polyacrylamide gels. A pI 4.1 endoribonuclease-containing 12S particle. B Affinity-purified U_1 snRNP. C Proteins eluted with 6M urea from an anti-U_1 snRNA-IgG column loaded with purified pI 4.1 endoribonuclease. *Arrows A*, *D*, *E*, and *F* mark the major bands; *arrows B* and *C* the minor ones; *arrows A1*, *A2*, and *A3*, further protein bands. The following protein standards were run in parallel gels: *1* bovine serum albumin (M_r = 68,000); *2* egg white albumin (M_r = 43,000); *3* chymotrypsinogen A (M_r = 25,000); *4* trypsin inhibition from soybean (M_r = 21,500) and *5* ribonuclease A (M_r = 13,700). (Bachmann et al. 1984a)

2. U1 snRNP-associated Poly(U)-specific Endoribonuclease

Based on evidence suggesting U1 snRNP's to be essential components of the mRNA splicing apparatus we searched for nuclease activity associated with U1 snRNP complexes. By an affinity chromatographic procedure using antibodies from a patient with systemic lupus erythematosus directed against U1 snRNP we succeeded in the isolation of an snRNP-associated poly(U)-degrading endoribonuclease (Bachmann et al. 1983a).

a) Purification of the Enzyme

The poly(U)-specific U1 snRNP-nuclease was prepared from rat liver nuclei by extraction and fractionation on a nonimmune IgG affinity column followed by immunoaffinity chromatography using immobilized IgG from a patient with systemic lupus erythematosus containing anti-U1 snRNP specificity and a second passage through the normal human IgG-Sepharose column (Bachmann et al. 1983a). The resulting purified RNP's consisted of four major and two minor polypeptides having mol. wts. of approx. 42,000, 13,500, 11,500, and 10,800 (major) and 29,000 and 26,000 (minor) (Fig. 10B). This polypeptide pattern is typical for that described for snRNP's (Lerner and Steitz 1979). The sedimentation coefficient of the isolated U snRNP obtained by sucrose gradient velocity centrifugation was determined to be 12S. Because the antibody preparation contained a high titer of anti-U1 snRNP the isolated snRNP should represent primarily U1 snRNP's.

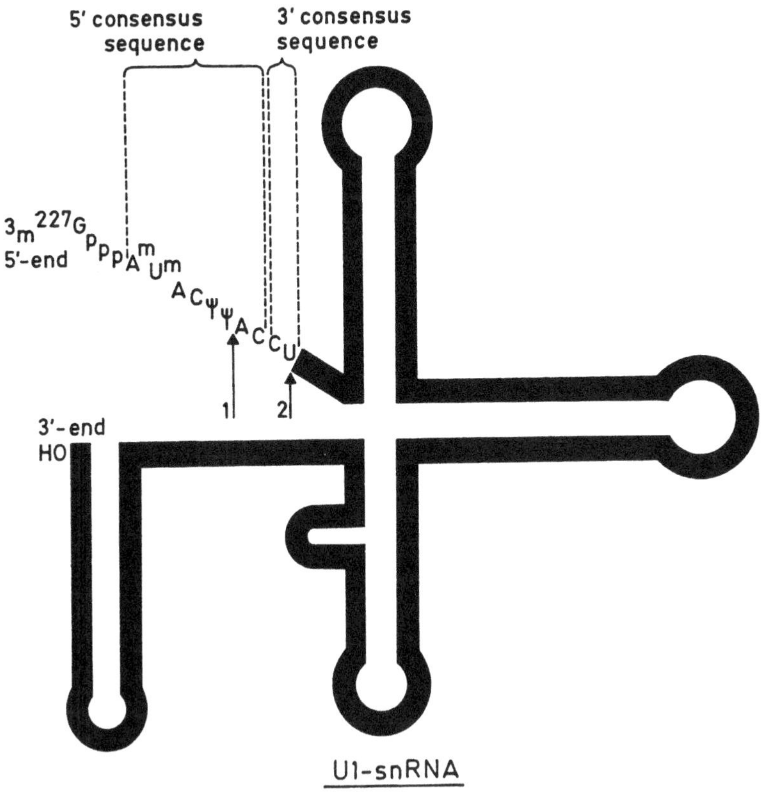

Fig. 11. Hypothesized modulation of function of U1 snRNP by base-specific splitting within the protein unprotected part of the RNA, catalyzed by U1 snRNP-associated poly(U)-specific endoribonuclease. Most of the U1 snRNA is protected against nucleolytic degradation by proteins as symbolized by the *thick black line*. *Arrows 1* and *2*, possible cleavage sites for the poly(U)-specific endoribonuclease. (Bachmann et al. 1984b)

b) Properties of the Enzyme

The U1 snRNP-associated endoribonuclease was found to be specific for poly(U). Other single-stranded polyribo- and polydeoxyribonucleotides, tRNA as well as double-stranded RNA and DNA were not degraded. The products of a terminal digestion are $(U)_{6-12}$ oligomers with 3'-OH and 5'-P end groups. The enzyme requires Mg^{2+} for optimal activity, possesses an pH optimum of 6.2 and is inhibited by N-ethylmaleimide. The Michaelis constant for poly(U) as substrate was 21 μM (Bachmann et al. 1983a).

c) Possible Function

Based on the observation that, firstly, antibodies to U1 snRNP inhibit splicing (Yang et al. 1981) and, secondly, snRNP's form complexes with hnRNP's (Prüsse et al. 1982), which are associated with the nuclear matrix during splicing (Mariman et al. 1982), it is proposed that snRNP's are provided with the enzyme machinery for cutting hnRNA's at their splicing sites (see Sect. E.I). At present, the function of the U1 snRNP-associated poly(U)-specific endoribonuclease is not yet clear. Perhaps the enzyme may remove oligo(U) sequences during processing of specialized, oligo(U)(+)hnRNA (Wood and Edmonds 1981). Another possible function may be deduced from the particular structure of the U1 snRNP complex. As shown in Figure 11, the greater part of the U1 snRNA within U1 snRNP particle is complexed with protein; only a short 5'terminal sequence of 11 nucleotide units is protein unprotected (Rinke et al. 1983). This nucleotide sequence which displays complementarity to the 3' and 5' donor and acceptor splice sites is thought to comprise the recognition site for the hnRNA splicing apparatus (Rogers and Wall 1980; Lerner et al. 1980). All but one uridylate residue in this short RNA segment are methylated or replaced by pseudouridines. It is not yet known whether the poly(U)-specific nuclease could cleave at pseudouridine residues; if so, one possible function of this enzyme might be the conversion of U1 to U1* (indicated by *arrow 1* in Fig. 11). Another uridylate residue at which the U1 snRNA might be cleaved by the poly(U)-specific nuclease is the eleventh nucleotide from the 5' end *(arrow 2)* which is the only one unmodified nucleotide that was found to be protein unprotected. Because by both reaction the interaction of the U1 snRNA with the complementary splice sites of hnRNA could be prevented, poly(U)-specific and U1 snRNP-associated endoribonuclease might attribute to a modulating function during splicing (Bachmann et al. 1984b).

3. 12S snRNP-associated pI 4.1 Endoribonuclease

During final purification of endoribonuclease VII by preparative isoelectric focusing (see Sect. E.II.1.a), a second endoribonuclease, called pI 4.1 endoribonuclease (Bachmann et al. 1984a), was separated from endoribonuclease VII (pI 8.5). This enzyme was found to be associated with 12S particles which copurify with the 45S RNP complex (Bachmann et al. 1983b). Experiments have shown that 75S - 30S RNP particles which are released from hnRNP particles by ribonuclease treatment (Prüsse et al. 1982; Pederson 1981) are associated with snRNP particles (Liautard et al. 1981) which are characterized by a sedimentation coefficient of 12S (Brunel et al. 1981). Moreover it is known that 12S snRNP and 30S hnRNP monomers can be isolated from a 45 - 50S RNP hypercomplex (Prüsse et al. 1982). The 12S snRNP in this complex contains U snRNA (R. Reddy and Busch 1981; Calvet and Pederson 1981). Recently, the 12S particles containing the pI 4.1 endonuclease could indeed be identified as U snRNP (Bachmann et al. 1984a). The properties of the pI 4.1 endoribonuclease are found to be very similar but not identical to the poly(U)-specific U1 snRNP-associated endoribonuclease.

a) Purification of the Enzyme

The purification scheme for the pI 4.1 endoribonuclease is analogous to that for endoribonuclease VII which has been described

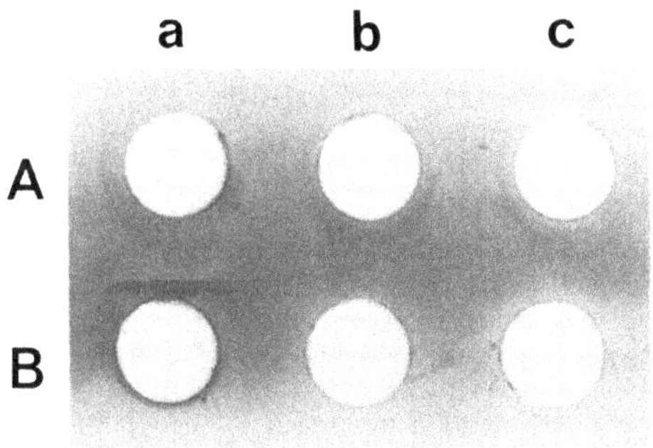

Fig. 12. Immunological characterization of the 12S RNP particles containing pI 4.1 endoribonuclease. The antigenicity of the 12S RNP complex isolated from calf thymus gland and found to be associated with the ribopyrimidine-specific pI 4.1 endoribonuclease (Bachmann et al. 1983b, 1984a) was checked by counterimmunoelectrophoresis (Kurata and Tan 1976). *Aa*, *Ab*, and *Ac* 12S RNP preparation; *Ba* serum containing anti-Sm; *Bb* control serum; *Bc* serum containing anti-RNP (Bachmann et al. 1984b)

above (Sect. E.II.1.a); the enzyme was obtained from calf thymus gland. During the final preparative isoelectric focusing step which had been performed in the pH range 3.5 - 10 in the presence of 4 M urea and dithiothreitol, the acidic pI 4.1 endoribonuclease was separated from the basic (pI 8.5) endoribonuclease VII (Bachmann et al. 1983b).

b) Properties of the Enzyme

The 12S RNP-associated pI 4.1 endoribonuclease hydrolyzes specifically poly(U) and poly(C) like as endoribonuclease VII; however, the intermediately formed oligomers are degraded up to 3'-UMP (85%) and P-2':3'-UMP (12%) (Bachmann et al. 1984a). For poly(U), the K_m value was determined to be 82 μM. The enzyme is an acidic protein (pI 4.1) and possesses an optimal pH 6.2. In contrast to endoribonuclease VII, the pyrimidine specific, acidic pI 4.1 endoriobnuclease is dependent on Mg^{2+} and is not stimulated by poly(A); the enzyme activity was found to be only strongly inhibited by this polymer. Moreover, pI 4.1 endoribonuclease is distinguished from endoribonuclease VII also by its resistance to thiol reagents and its low sensitivity towards EDTA. 0.1 M NaCl abolished the enzyme activity by about 95%. No binding of poly(A) to the enzyme could be detected using the nitrocellulose filter technique. The pI 4.1 endoribonuclease was found to be associated with 12S particles that can be characterized as snRNP's by several criteria; these include the protein pattern obtained in sodium dodecyl sulfate polyacrylamide gels, which is typical for that of immunoaffinity purified U snRNP's (Lerner and Steitz 1979; White et al. 1981; Zeller et al. 1983; Fig. 10), and the antigenicity of these particles as checked by counterimmunoelectrophoresis (Fig. 12). In sodium dodecyl sulfate polyacrylamide gels the potential Sm antigens D (M_r = 13,500), E (M_r =11,500) and F (M_r = 10,800) (Lerner and Steitz 1979) were found to be present in the 12S particles (Fig. 10). In addition to these polypeptides some further protein bands (A, M_r = 42,000; A1, M_r = 47,000; A2, M_r = 37,000; A3, M_r = 33,000; B, M_r = 29,000 and C, M_r = 26,000) were found, which could be partially (A, A1 and A2) adsorbed to an anti-U1 snRNP IgG column. Although the pI 4.1 endoribonuclease shows striking similarities with the snRNP-associated poly(U)-specific endoribonuclease (association with 12S particles, Mg^{2+} requirement, pH optimum of 6.2), the latter enzyme is distinguished from the pI 4.1 nuclease by the fact, that it hydrolyzes exclusively poly(U) creating oligo(U) fragments with 3'-OH and 5'-P termini.

c) Possible Function

At present, the functional significance of the 12S RNP-associated pI 4.1 nuclease is not known. The experimental data suggest that the 12S RNP particles containing the enzyme are partially related to the U1 snRNP's. Therefore, we hypothesize that this nuclease is involved in the inactivation of U1 snRNA function during hnRNA splicing by converting U1 snRNA to U1* snRNA which has lost its property to associate with hnRNA (Lerner et al. 1980; Greer et al. 1983).

F. Role of Poly(A) Chain Length Controlled by Specific Nucleases and of snRNP's during Poly(A)(+)mRNA-Nuclear Matrix Association and Nucleocytoplasmic Poly(A)(+)mRNA Transport

Within the nucleus, the hnRNA is found to be almost completely associated with the nuclear matrix (T.E. Miller et al. 1978; Herman et al. 1978; Van Eekelen and Van Venrooij 1981). The nuclear matrix seems to provide the platform for the posttranscriptional processing reactions such as cap formation (Jackson et al. 1981), polyadenylation and splicing (Bouteille et al. 1983). The 3'terminal poly(A) sequence of hnRNA and mRNA is thought to be involved in attachment of the nuclear matrix (Herman et al. 1978). Perhaps the P73 poly(A)-associated protein, which had been identified in the nuclear matrix of duck erythroblasts (Maundrell et al. 1981), may mediate RNA•matrix association. Moreover, snRNP's have been shown to be nearly quantitatively associated with the nuclear matrix (T.E. Miller et al. 1978; Maundrell et al. 1981). In one model it was proposed (R. Reddy and Busch 1983) that snRNP's function during transport of hnRNP to the nuclear pore complexes as connecting links between the information-bearing macromolecules and the nuclear matrix elements.

Nucleocytoplasmic transport of poly(A)(+)mRNA is an ATP-dependent process which is thought to be mediated by a nuclear envelope nucleoside triphosphatase (NTPase; Agutter et al. 1976, 1979; Müller et al. 1984). The activity of this enzyme was found to be stimulated by RNA (Agutter 1977). Bernd et al. (1982b) demonstrated that synthetic poly(A) or natural poly(A)(+)mRNA are by far the most potent stimulators of NTPase activity; stimulation by poly(A), however, provides a minimal chain length of 18 adenyate residues. In contrast, poly(A)(-)mRNA which had been obtained by selective removal of the poly(A) tail of mRNA by endoribonuclease IV, had no stimulatory effect on NTPase activity (Bernd et al. 1982b). Therefore we conclude that the synthesis of the 3'terminal poly(A) sequence of mRNA is a prerequisite for nucleocytoplasmic translocation of this RNA species. This assumption is supported by the recent findings (Villarreal and White 1983) that nuclear RNA transcripts from a late region deletion mutant of SV40, which cannot be polyadenylated, are not transported out of the nucleus. As consequence, poly(A) nucleases have to play a crucial role in determining whether or not a defined mRNA species is transported through the nuclear pore complex. On the other hand, histone mRNA's which lack poly(A) are found to be efficiently transported to the cytoplasm. Moreover, adenovirus type 2 poly(A)(-)mRNA synthesized in 3'-deoxyadenosine-treated cells has shown to be transported at nearly the same rate as the corresponding poly(A)(+)mRNA (Zeevi et al. 1982).

In the cytoplasm the poly(A)(+)mRNA is found again to be attached to cytoskeletal structures. Thus, the polyribosomal complex has been shown to be fixed to the cytoskeleton via its mRNA component (Lenk et al. 1977; Van Venrooij et al. 1981). Milcarek and Penman (1974) proposed the P76 poly(A)-associated protein to be involved in attaching polyribosomes to the cytoskeleton. Jeffery (1982) suggested that the binding of the cytoplasmic mRNA to the cytoskeletal framework provides the eukaryotic cell with a mechanism for controlling the polarity of mRNA distribution. Thus Phillips (1982) demonstrated an uneven regional distribution of poly(A) within *Xenopus* embryos, which is altered during early development. Moreover, microinjection experiments of poly(A)(+)mRNA into fertilized *Xenopus* eggs suggested that specific kinds of mRNA are selectively concentrated in different regions of the cell (Capco and Jeffery 1981). In this context, it may be important that actin as well as microtubuli protein are able to reduce the activities of the poly(A) anabolic poly(A) polymerase as well as of the poly(A) catabolic endoribonuclease IV (Schröder et al. 1982b). Moreover, poly(A) has been demonstrated to display a high affinity to microtubule-associated tau proteins (Schröder et al. 1984a), provided that its chain length is greater than 12 nucleotide units (Schröder et al. 1982a). Thus poly(A)-specific ribonucleases, controlling the size of the 3'terminal poly(A) segment of hnRNA and mRNA in cooperation with poly(A) polymerases and poly(A)-associated proteins, may play an essential part during intracellular compartmentalization and realization of the genetic information.

G. Conclusion and Future Directions

During maturation of hnRNA to mRNA the size of the primary transcript is altered by a series of enzymic processes including both excision of internal noncoding sequences and cleavage/polyadenylation and polydeadenylation events at the 3'terminus. Therefore, nucleotide-specific ribonucleases may play an essential role during the processing reactions. In contrast to the 3' terminal cleavage and internal cleavage/ligation processes, the molecular basis of the enzyme and protein controlled 3'poly(A) chain length regulation is known in first outlines. Two specific poly(A) splitting enzymes, the endoribonuclease IV and the poly(A)-specific 2',3'-exoribonuclease, have been shown to be implicated in this reaction. The activity of endoribonuclease IV is controlled by posttranscriptional phosphorylation as well as by specific poly(A)-associated proteins and cytoskeletal elements. On the other hand, the poly(A)-specific 2',3'-exoribonuclease has been shown to be involved in 2',5'-oligo(A) metabolism, too. Concerning RNA:RNA splicing, three novel enzymes are described, which are potential candidates for the excision of the introns or for the modulation of snRNP function. Thus, endoribonuclease VII has been shown to be capable of recognizing internal oligo(U) sequences in the mRNA precursor and to be modulated by the 3'terminal poly(A) sequence of mRNA. U1 snRNP's were found to contain a poly(U)-specific endoribonuclease probably involved in U1 → U1* conversion controlling U1 snRNP - hnRNP interaction. Nevertheless, the function of these enzymes during RNA splicing remains to be established. Moreover, the elucidation of the 3'end formation of hnRNA by the cleavage/polyadenylation step may be a task for the future. It would be particularly interesting to know if snRNP's or snRNP-associated endoribonucleases are involved in this step.

Acknowledgement. This work was supported by a grant from the Deutsche Forschungsgemeinschaft (Mu 348/7-4). We appreciate the helpful assistance of Ms. K. Bartelt. H.C.S. was a recipient of a Liebig stipendium from the Fonds der Chemischen Industrie.

References

Abelson, J.: RNA processing and the intervening sequence problem. Annu. Rev. Biochem. 48, 1035-1069 (1979)

Abraham, A.K., Jacob, S.T.: Hydrolysis of poly(A) to adenine nucleotides by purified poly(A) polymerase. Proc. Natl. Acad. Sci. USA 75, 2085-2087 (1978)

Adams, D.S., Noonan, D., Jeffery, W.R.: The poly(adenylic acid) · protein complex is restricted to the nonpolysomal messenger ribonucleoprotein of Physarum polycephalum. Biochemistry 19, 1965-1970 (1980)

Adesnik, M., Darnell, J.E.: Biogenesis and characterization of histone messenger RNA in HeLa cells. J. Mol. Biol. 67, 397-406 (1972)

Agutter, P.S., McArdle, H.J., McCaldin, B.: Evidence for involvement of nuclear envelope nucleoside triphosphatase in nucleocytoplasmic translocation of ribonucleoprotein. Nature (London) 263, 165-167 (1976)

Agutter, P.S., Harris, J.H., Stevenson, I.: Ribonucleic acid stimulation of mammalian liver nuclear-envelope nucleoside triphosphatase. Biochem. J. 162, 671-679 (1977)

Agutter, P.S., McCaldin, B., McArdle, H.J.: Importance of mammalian nuclear-envelope nucleoside triphosphatase in nucleo-cytoplasmic transport of ribonucleoproteins. Biochem. J. 182, 811-819 (1979)

Altman, S., Guerrier-Takada, C., Frankfort, H.M., Robertson, H.D.: RNA-processing nucleases. In: Nucleases (eds. S.M. Linn, R.J. Roberts), pp. 243-274. New York: Cold Spring Harbor 1982

Amara, S.G., Jonas, V., Rosenfeld, M.G., Ong, E.S., Evans, R.M.: Alternative RNA processing in calcitonin gene expression generates mRNAs encoding different polypeptide products. Nature (London) 298, 240-244 (1982)

Babich, A., Nevins, J.R., Darnell, J.E.: Early capping of transcripts from the adenovirus major late transcription unit. Nature (London) 287, 246-248 (1980)

Bachmann, M., Trautmann, F., Messer, R., Zahn, R.K., Meyer zum Büschenfelde, K.-H., Müller, W.E.G.: Association of a polyuridylate-specific endoribonuclease with small nuclear ribonucleo-proteins which had been isolated by affinity chromatography using antibodies from a patient with systemic lupus erythematosus. Eur. J. Biochem. 136, 447-451 (1983a)

Bachmann, M., Zahn, R.K., Müller, W.E.G.: Purification and properties of a novel pyrimidine-specific endoribonuclease termed endoribonuclease VII from calf thymus that is modulated by polyadenylate. J. Biol. Chem. 258, 7033-7040 (1983b)

Bachmann, M., Messer, R., Trautmann, F., Müller, W.E.G.: 12S small nuclear ribonucleoprotein-associated acidic and pyrimidine-specific endoribonuclease from calf thymus and L5178y cells. Biochim. Biophys. Acta 783, 89-99 (1984a)

Bachmann, M., Schröder, H.C., Messer, R., Müller, W.E.G.: Base-specific ribonucleases potentially involved in heterogeneous nuclear RNA processing and poly(A) metabolism. FEBS Lett. 171, 25-30 (1984b)

Baer, B.W., Kornberg, R.D.: Repeating structure of cytoplasmic poly(A) ribonucleoprotein. Proc. Natl. Acad. Sci. USA 77, 1890-1892 (1980)

Baglioni, C., Maroney, P.A., West, D.K.: 2'5'oligo(A) polymerase activity and inhibition of viral RNA synthesis in interferon-treated HeLa cells. Biochemistry 18, 1765-1770 (1979)

Ball, L.A.: Induction of 2'5'-oligoadenylate synthetase activity and a new protein by chick interferon. Virology 94, 282-296 (1979)

Bard, E., Efron, D., Marcus, A., Perry, R.P.: Translational capacity of deadenylated RNA. Cell 1, 101-106 (1974)

Baydoun, H., Hoppe, J., Freist, W., Wagner, K.G.: Purification and ATP substrate site of a cyclic nucleotide-independent protein kinase (NI) from porcine liver nuclei. J. Biol. Chem. 257, 1032-1036 (1982)

Benoist, C., O'Hare, K., Breathnach, R., Chambon, P.: The ovalbumin gene-sequence of putative control regions. Nucleic Acids Res. 8, 127-142 (1980)

Berget, S.M.: Are U4 small nuclear ribonucleoproteins involved in polyadenylation? Nature (London) 309, 179-182 (1984)

Bergmann, J.E., Brawerman, G.: Control of breakdown of the polyadenylate sequence in mammalian polyribosomes: role of poly(adenylic acid)-protein interactions. Biochemistry 16, 259-264 (1977)

Bernardi, A., Bernardi G.: Studies on acid hydrolases. III. Isolation and properties of spleen acid ribonuclease. Biochim. Biophys. Acta 129, 23-31 (1966)

Bernd, A., Batke, E., Zahn, R.K., Müller, W.E.G.: Age-dependent gene induction in quail oviduct. XV. Alterations of the poly(A)-associated protein pattern and of the poly(A) chain length of mRNA. Mech. Ageing Dev. 19, 361-377 (1982a)

Bernd, A., Schröder, H.C., Zahn, R.K., Müller, W.E.G.: Modulation of the nuclear-envelope nucleoside triphosphatase by poly(A)-rich mRNA and by microtubule-protein. Eur. J. Biochem. 129, 43-49 (1982b)

Bernd, A., Zahn, R.K., Maidhof, A., Müller, W.E.G.: Analysis of polyadenylate-protein complex of polysomal messenger RNA from mouse L cells. Hoppe-Seyler's Z. Physiol. Chem. 363, 221-228 (1982c)

Bina, M., Feldmann, R.J., Deeley, R.G.: Could poly(A) align the splicing sites of messenger RNA precursors? Proc. Natl. Acad. Sci. USA 77, 1278-1282 (1980)

Birchmeier, C., Grosschedl, R., Birnstiel, M.L.: Generation of authentic 3' termini of an H2A mRNA in vivo is dependent on a short inverted DNA repeat and on spacer sequences. Cell 28, 739-745 (1982)

Blobel, G.: A protein of molecular weight 78,000 bound to the polyadenylate region of eukaryotic messenger RNAs. Proc. Natl. Acad. Sci. USA 70, 924-928 (1973)

Bothwell, A.L.M., Altman, S.: Partial purification and properties of an endoribonuclease isolated from human KB cells. J. Biol. Chem. 250, 1451-1459 (1975)

Bouteille, M., Bouvier, D., Seve, A.P.: Heterogeneity and territorial organization of the nuclear matrix and related structures. Int. Rev. Cytol. 83, 135-182 (1983)

Brawerman, G.: Alterations in the size of the poly(A) segment in newly-synthesized messenger RNA of mouse sarcoma 180 ascites cells. Mol. Biol. Rep. 1, 7-13 (1973)

Brawerman, G.: Eukaryotic messenger RNA. Annu. Rev. Biochem. 43, 621-642 (1974)

Brawerman, G.: Characteristics and significance of the poly adenylate sequence in mammalian messenger RNA. Prog. Nucleic Acid Res. Mol. Biol. 17, 117-148 (1976)

Brunel, C., Widada, J.S., Lelay, M.N., Jeanteur, P., Liautard, J.P.: Purification and characterization of a simple ribonucleoprotein particle containing small nucleoplasmic RNAs (snRNP) as a subset of RNP containing heterogenous nuclear RNA (hnRNP) from HeLa cells. Nucleic Acids Res. 9, 815-830 (1981)

Brunel, C., Widada, J.S., Jeanteur, P.: snRNPs and scRNPs in eukaryotic cells. In: Progress in Molecular and Subcellular Biology (eds. F.E. Hahn, D.J. Kopecko, W.E.G. Müller), Vol. 9, pp. 1-52. Berlin-Heidelberg-New York: Springer 1985

Bryan, J., Nagle, B.W., Doenges, K.H.: Inhibition of tubulin assembly by RNA and other polyanions: evidence for a required protein. Proc. Natl. Acad. Sci. USA 72, 3570-3574 (1975)

Busch, H., Reddy, R., Rothblum, L., Choi, Y.C.: SnRNAs, snRNPs, and RNA processing. Annu. Rev. Biochem. 51, 617-654 (1982)

Calvet, J.P., Pederson, T.: Base-pairing interactions between small nuclear RNAs and nuclear RNA precursors as revealed by psoralen cross-linking in vivo. Cell 26, 363-370 (1981)

Capco, D.G., Jeffery, W.R.: Regional accumulation of vegetal pole poly(A)$^{+}$RNA injected into fertilized eggs of Xenopus laevis. Nature (London) 294, 255-257 (1981)

Carlin, R.K.: The poly(A) segment of mRNA: (1) Evolution and function and (2) The evolution of viruses. J. Theor. Biol. 71, 323-338 (1978)

Cech, T.R.: RNA splicing: three themes with variations. Cell 34, 713-716 (1983)

Chaudhari, N., Hahn, W.E.: Genetic expression in the developing brain. Science 220, 924-928 (1983)

Chen-Kiang, S., Nevins, J.R., Darnell, J.E.: N-6 methyl-adenosine in adenovirus type 2 nuclear RNA is conserved in the formation of messenger RNA. J. Mol. Biol. 135, 733-752 (1979)

Chikaraishi, D.M., Deeb, S.S., Sueoka, N.: Sequence complexity of nuclear RNAs in adult rat tissues. Cell 13, 111-120 (1978)

Ciejek, E.M., Tsai, M.-J., O'Malley, B.W.: Actively transcribed genes are associated with the nuclear matrix. Nature (London) 306, 607-609 (1983)

Darnell, J.E.: Steps in processing of mRNA: implications for gene regulation. In: From Gene to Protein: Information Transfer in Normal and Abnormal Cells (eds. T.R. Russell, K. Brew, H. Faber, J. Schultz), pp. 207-228. London-New York: Academic Press 1979

Darnell, J.E.: Variety in the level of gene control in eukaryotic cells. Nature (London) 297, 365-371 (1982)

Darnell, J.E., Jelinek, W.R., Molloy, G.R.: Biogenesis of mRNA. Genetic regulation in mammalian cells. Science 181, 1215-1221 (1973)

Dunn, J.J., Studier, F.J.: T7 early RNAs and Escherichia coli ribosomal RNAs are cut from large precursors RNAs in vivo by ribonuclease III. Proc. Natl. Acad. Sci. USA 70, 3296-3300 (1973)

Economidis, I.V., Pederson, T.: Structure of nuclear ribonucleoprotein: Heterogeneous nuclear RNA is complexed with a major sextet of proteins in vivo. Proc. Natl. Acad. Sci. USA 80, 1599-1602 (1983)

Edmonds, M.: Poly(A) adding enzymes. In: The Enzymes (ed. P.D. Boyer), Vol. XV, pp. 217-244. London-New York: Academic Press 1982

Edmonds, M., Abrams, R.: Polynucleotide biosynthesis: Formation of a sequence of adenylate units from adenosine triphosphate by an enzyme from thymus nuclei. J. Biol. Chem. 235, 1142-1149 (1960)

Edmonds, M., Winters, M.A.: Polyadenylate polymerases. Prog. Nucleic Acid Res. Mol. Biol. 17, 149-179 (1976)

Edmonds, M., Nakazato, H., Korwek, E.L., Venkatesan, S.: Transcribed oligonucleotide sequences in HeLa cell hnRNA and mRNA. Prog. Nucleic Acid Res. Mol. Biol. 19, 99-112 (1976)

Eichler, D.C., Eales, S.J.: Isolation and characterization of a single-stranded specific endoribonuclease from Ehrlich cell nucleoli. J. Biol. Chem. 257, 14384-14389 (1982)

Eppstein, D.A., Barnett, J.W., Marsh, Y.V., Gosselin, G., Imbach, J.-L.: Xyloadenosine analogue of $(A2'p)_2A$ inhibits replication of herpes simplex virus 1 and 2. Nature (London) 302, 723-724 (1983)

Esumi, H., Takahashi, Y, Sekiya, T., Sato, S., Nagase, S., Sugimura, T.: Presence of albumin mRNA precursors in nuclei of analbuminemic rat liver lacking cytoplasmic albumin mRNA. Proc. Natl. Acad. Sci. USA 79, 734-738 (1982)

Falkenthal, S., Graham, M.L., Korn, E.L., Lengyel, J.A.: Transcription, processing and turnover of RNA from the Drosophila mobile genetic element copia. Dev. Biol. 92, 294-305 (1982)

Fitzgerald, M., Shenk, T.: The sequence 5'-AAUAAA-3' forms part of the recognition site for polyadenylation of late SV40 mRNAs. Cell 24, 251-260 (1981)

Floyd-Smith, G., Slattery, E., Lengyel, P.: Interferon action: RNA cleavage pattern of a (2'-5')oligoadenylate-dependent endonuclease. Science 212, 1030-1032 (1981)

Frank, J.J., Levy, C.C.: Properties of a human liver ribonuclease. J. Biol. Chem. 251, 5745-5751 (1976)

Galli, G., Hofstetter, H., Stunnenberg, H.G., Birnstiel, M.L.: Biochemical complementation with RNA in the Xenopus oocyte: a small RNA is required for the generation of 3' histone mRNA termini. Cell 34, 823-828 (1983)

Gallinaro, H., Lazar, E., Jacob, M., Krol, A., Branlant, C.: Small RNAs in hnRNP fibrils and their possible function in splicing. Mol. Biol. Rep. 7, 31-39 (1981)

Gosselin, G., Imbach, J.L.: Synthese du trimere de la β-D-xylofuranosyl-9 adenine a liaisons internucleotidiques 2'→5'. Tetrahedron Lett. 22, 4699-4702 (1981)

Green, M.R., Maniatis, T., Melton, D.A.: Human β-globin pre-mRNA synthesized in vitro is accurately spliced in Xenopus oocyte nuclei. Cell 32, 681-694 (1983)

Greer, C.L., Peebles, C.L., Gegenheimer, P., Abelson, J.: Mechanism of action of a yeast RNA ligase in tRNA splicing. Cell 32, 537-546 (1983)

Guha, A.: An exoribonuclease in bovine brain. Brain Res. 83, 65-79 (1975)

Guha, A.: Purification and mode of action of exoribonuclease from bovine brain. J. Biol. Chem. 252, 6416-6420 (1977)

Harada, F., Kato, N.: Nucleotide sequences of 4.5S RNAs associated with poly(A)-containing RNAs of mouse and hamster cells. Nucleic Acids Res. 8, 1273-1285 (1980)

Hausen, P., Stein, H.: Ribonuclease H. An enzyme degrading the RNA moiety of DNA-RNA hybrids. Eur. J. Biochem. 14, 278-283 (1970)

Hendrick, J.P., Wolin, S.L., Rinke, J., Lerner, M.R., Steitz, J.A.: Ro small cytoplasmic ribonucleoproteins are a subclass of La ribonucleoproteins: further characterization of the Ro and La small ribonucleoproteins from uninfected mammalian cells. Mol. Cell. Biol. 1, 1138-1149 (1981)

Hentschel, C.C., Birnstiel, M.L.: The organization and expression of histone gene families. Cell 25, 301-313 (1981)

Herman, R., Weymouth, L., Penman, S.: Heterogeneous nuclear RNA-protein fibers in chromatin-depleted nuclei. J. Cell Biol. 78, 663-674 (1978)

Higgs, D.R., Goodbourn, S.E.Y., Lamb, J., Clegg, J.B., Weatherall, D.J.: α-Thalassaemia caused by a polyadenylation signal mutation. Nature (London) 306, 398-400 (1983)

Hovanessian, A.G., Brown, R.E., Kerr, I.M.: Synthesis of low molecular weight inhibitor of protein synthesis with enzyme from interferon-treated cells. Nature (London) 268, 537-540 (1977)

Howard, E.F.: Small nuclear RNA molecules in nuclear ribonucleoprotein complexes from mouse erythroleukemia cells. Biochemistry 17, 3228-3236 (1978)

Imaizumi-Scherrer, M.-T., Maundrell, K., Civelli, O., Scherrer, K.: Transcriptional and post-transcriptional regulation in duck erythroblasts. Dev. Biol. 93, 126-138 (1982)

Ishikawa, K., Sato-Odani, S., Ogata, T.: The role of ATP in the transport of rapidly-labeled RNA from isolated nuclei of rat liver in vitro. Biochim. Biophys. Acta 521, 650-661 (1978)

Ittel, M.E., Niedergang, C., Munoz, D., Petek, F., Okazaki, H., Mandel, P.: Partial purification and characterization of an acid ribonuclease in beef brain nuclei. J. Neurochem. 25, 171-176 (1975)

Jackson, D.A., McCready, S.J., Cook, P.R.: RNA is synthesized at the nuclear cage. Nature (London) 292, 552-555 (1981)

Jacob, S.T., Rose, K.M.: RNA polymerases and poly(A) polymerase from neoplastic tissues and cells. In: Methods in Cancer Research (ed. H. Busch), Vol. XIV, pp. 191-241. London-New York: Academic Press 1978

Jacobson, A., Favreau, M.: Possible involvement of poly(A) in protein synthesis. Nucleic Acids Res. 11, 6353-6368 (1983)

Jeffery, W.R.: Characterization of polypeptides associated with messenger RNA and its polyadenylate segment in Ehrlich ascites messenger ribonucleoprotein. J. Biol. Chem. 252, 3535-3532 (1977)

Jeffery, W.R.: Composition and properties of messenger ribonucleoprotein fragments containing and lacking polyadenylate. Biochim. Biophys. Acta 521, 217-228 (1978)

Jeffery, W.R.: Messenger RNA in the cytoskeletal framework: analysis by in situ hybridization. J. Cell Biol. 95, 1-7 (1982)

Jeffreys, A.J., Flavell, R.A.: The rabbit β-globin gene contains an large insert in the coding sequence. Cell 12, 1097-1108 (1977)

Kates, J., Beeson, J.: Ribonucleic acid synthesis in vaccinia virus. II. Synthesis of polyriboadenylic acid. J. Mol. Biol. 50, 19-33 (1970)

Kerr, I.M., Brown, R.E.: pppA2'p5'A2'p5'A: an inhibitor of protein synthesis synthesized with an enzyme fraction from interferon-treated cells. Proc. Natl. Acad. Sci USA 75, 256-260 (1978)

Khasigov, P.Z., Glazkov, V.F., Del'Vig, A.A., Kuznetsov, D.A., Nikolaev, A.Y.: Age changes in the metabolism of nuclear procursors of mRNA in liver and brain cortex cells of the rat. Biokhimiya 48, 179-185 (1983)

King, C.R., Piatigorsky, J.: Alternative RNA splicing of the murine αA-crystallin gene: protein-coding information within an intron. Cell 32, 707-712 (1983)

Kish, V.M., Pederson, T.: Ribonucleoprotein organization of polyadenylate sequences in HeLa cell heterogeneous nuclear RNA. J. Mol. Biol. 95, 227-238 (1975)
Kish, V.M., Pederson, T.: Heterogeneous nuclear RNA secondary structure: oligo(U) sequences base-paired with poly(A) and their possible role as binding sites for heterogeneous nuclear RNA-specific proteins. Proc. Natl. Acad. Sci. USA 74, 1426-1430 (1977)
Kole, R., Altman, S.: Properties of purified ribonuclease P from Escherichia coli. Biochemistry 20, 1902-1906 (1981)
Korwek, E.L., Nakazato, H., Venkatesan, S., Edmonds, M.: Poly(uridylic acid) sequences in messenger ribonucleic acid of HeLa cells. Biochemistry 15, 4643-4649 (1976)
Kouidou, S., Triantos, A., Kavoukopoulos, E., Trakatellis, A.: Endoplasmic reticulum nuclease. Purification and specificity. Eur. J. Biochem. 120, 9-14 (1981)
Krieg, P.A., Melton, D.A.: Formation of the 3' end of histone mRNA by post-transcriptional processing. Nature (London) 308, 203-206 (1984)
Kruger, K., Grabowski, P.J., Zaug, A.J., Sands, J., Gottschling, D.E., Cech, T.R.: Self-splicing RNA: autoexcision and autocyclization of the ribosomal RNA intervening sequence of Tetrahymena. Cell 31, 147-157 (1982)
Kumagai, H., Igarashi, K., Takayama, T., Watanabe, K., Sugimoto, K., Hirose, S.: A microsomal endoribonuclease from rat liver. Biochim. Biophys. Acta 608, 324-331 (1980)
Kurata, N., Tan, E.M.: Identification of antibodies to nuclear acidic antigens by counterimmunoelectrophoresis. Arthritis Rheum. 19, 574-580 (1976)
Kwan, C.N.: Purification and characterization of an endoribonuclease from nucleoplasm and nucleoli of HeLa cells. J. Biol. Chem. 251, 7132-7136 (1976)
Kwan, C.N.: A cytoplasmic exoribonuclease from HeLa cells. Biochim. Biophys. Acta 479, 322-331 (1977)
Kwan, S.W., Brawerman, G.: A particle associated with the polyadenylate segment in mammalian messenger RNA. Proc. Natl. Acad. Sci. USA 69, 3247-3250 (1972)
Lazarus, H.M., Sporn, M.B.: Purification and properties of a nuclear exoribonuclease from Ehrlich ascites tumor cells. Proc. Natl. Acad. Sci. USA 57, 1386-1393 (1967)
Lenk, R., Ransom, L., Kaufmann, Y., Penman, S.: A cytoskeletal structure with associated polyribosomes obtained from HeLa cells. Cell 10, 67-68 (1977)
Lerner, M.R., Steitz, J.A.: Antibodies to small nuclear RNAs complexed with proteins are produced by patients with systemic lupus erythematosus. Proc. Natl. Acad. Sci. USA 76, 5495-5499 (1979)
Lerner, M.R., Steitz, J.A.: Snurps and scyrps. Cell 25, 298-300 (1981)
Lerner, M.R., Boyle, J.A., Mount, S.M., Wolin, S.L., Steitz, J.A.: Are snRNPs involved in splicing? Nature (London) 283, 220-221 (1980)
Lerner, M.R., Boyle, J.A., Hardin, J.A., Steitz, J.A.: Two novel classes of small ribonucleoproteins detected by antibodies associated with lupus erythematosus. Science 211, 400-402 (1981)
Levenson, R.G., Marcu, K.B.: On the existence of polyadenylated histone mRNA in Xenopus laevis oocytes. Cell 9, 311-322 (1976)
Levy, C.C., Karpetsky, T.P.: The purification and properties of chicken liver RNase. An enzyme which is useful in distinguishing between cytidylic and uridylic acid residues. J. Biol. Chem. 255, 2153-2159 (1980)
Levy, C.C., Schmuckler, M., Frank, J.J., Karpetsky, T.P., Jewett, P.B., Hieter, P.A., Legendre, S.M., Dorr, R.G.: Possible role for poly(A) as an inhibitor of endonuclease activity in eukaryotic cells. Nature (London) 256, 340-342 (1975)
Liautard, J.P., Sri Widada, J., Brunel, C.: Particles containing small molecular weight nuclear RNAs (snRNPs). Structure and possible function. Mol. Biol. Rep. 7, 41-45 (1981)

Littauer, U.Z., Soreq, H.: The regulatory function of poly(A) and adjacent 3' sequences in translated RNA. Prog. Nucleic Acid Res. Mol. Biol. 27, 53-83 (1982)

Long, B.H., Huang, C.-Y., Pogo, A.O.: Isolation and characterization of the nuclear matrix in Friend erythroleukemia cells: chromatin and heterogeneous RNA interactions with the nuclear matrix. Cell 18, 1079-1090 (1979)

Mangiarotti, G., Zuker, C., Chisholm, R.L., Lodish, H.F.: Different mRNAs have different nuclear transit times in Dictyostelium discoideum aggregates. Mol. Cell. Biol. 3, 1511-1517 (1983)

Manley, J.L.: Accurate and specific polyadenylation of mRNA precursors in a soluble whole-cell lysate. Cell 33, 595-605 (1983a)

Manley, J.L.: Analysis of the expression of genes encoding animal mRNA by in vitro techniques. Prog. Nucleic Acid Res. Mol. Biol. 30, 195-244 (1983b)

Marbaix, G., Huez, G., Soreq, H.: What is the role of poly A on eukaryotic messengers? Trends Biochem. Sci. 2, N106-N107 (1977)

Mariman, E.C.M., Van Eekelen, C.A.G., Reinders, R.J., Berns, A.J.M., Van Venrooij, W.J.: Adenoviral heterogeneous nuclear RNA is associated with the host nuclear matrix during splicing. J. Mol. Biol. 154, 103-119 (1982)

Matts, R.L., Siegel, F.L.: Regulation of hepatic poly(A) endonuclease by corticosterone and amino acids. J. Biol. Chem. 254, 11228-11233 (1979)

Maundrell, K., Maxwell, E.S., Puvion, E., Scherrer, K.: The nuclear matrix of duck erythroblasts is associated with globin mRNA coding sequences but not with the major proteins of 40S nuclear RNP. Exp. Cell Res. 136, 435-445 (1981)

Melton, D.A., DeRobertis, E.M., Cortese, R.: Order and intracellular location of the events involved in the maturation of a spliced RNA. Nature (London) 284, 143-148 (1980)

Meyuhas, O., Perry, R.P.: Relationship between size, stability and abundance of the messenger RNA of mouse L cells. Cell 16, 139-148 (1979)

Milcarek, C., Penman, S.: Membrane-bound polyribosomes in HeLa cells: association of polyadenylic acid with membranes. J. Mol. Biol. 89, 327-338 (1974)

Miller, T.E., Huang, C.Y., Pogo, A.O.: Rat liver nuclear skeleton and ribonucleoprotein complexes containing hnRNA. J. Cell Biol. 76, 675-691 (1978)

Miller, H.I., Riggs, A.D., Gill, G.N.: Ribonuclease H (hybrid) in Escherichia coli. Identification and characterization. J. Biol. Chem. 248, 2621-2624 (1973)

Milner, R.J., Bloom, F.E., Lai, C., Lerner, R.A., Sutcliffe, J.G.: Brain-specific genes have identifier sequences in their introns. Proc. Natl. Acad. Sci. USA 81, 713-717 (1984)

Minks, M.A., Benvin, S., Maroney, P.A., Baglioni, C.: Metabolic stability of 2'5'-oligo(A) and activity of 2'5'oligo(A)-dependent endonuclease in extracts of control and interferon-treated HeLa cells. Nucleic Acids Res. 6, 767-780 (1979)

Molloy, G.R., Jelinek, W., Salditt, M., Darnell, J.E.: Arrangement of specific oligonucleotides within poly(A) terminated hnRNA molecules. Cell 1, 43-53 (1974)

Molloy, G.R., Thomas, W.L., Darnell, J.E.: Occurrence of uridylate-rich oligonucleotide regions in heterogeneous nuclear RNA of HeLa cells. Proc. Natl. Acad. Sci. USA 69, 3684-3688 (1972)

Montell, C., Fisher, E.F., Caruthers, M.H., Berk, A.J.: Inhibition of RNA cleavage but not polyadenylation by a point mutation in mRNA 3' consensus sequence AAUAAA. Nature (London) 305, 600-605 (1983)

Mount, S.M.: A catalogue of splice junction sequences. Nucleic Acids Res. 10, 453-472 (1982)

Mount, S.M., Pettersson, I., Hinterberger, M., Karmas, A., Steitz, J.A.: The U1 small nuclear RNA-protein complex selectively binds a 5' splice site in vitro. Cell 33, 509-518 (1983)

Müller, W.E.G.: Endoribonuclease IV. A poly(A)-specific ribonuclease from chick oviduct. 1. Purification of the enzyme. Eur. J. Biochem. 70, 241-248 (1976)
Müller, W.E.G., Totsuka, A., Kroll, I., Nusser, I., Zahn, R.K.: Poly(A) polymerase in quail oviduct. Changes during estrogen induction. Biochim. Biophys. Acta 383, 147-159 (1975)
Müller, W.E.G., Seibert, G., Steffen, R., Zahn, R.K.: Endoribonuclease IV, 2. Further investigation on the specificity. Eur. J. Biochem. 70, 249-258 (1976)
Müller, W.E.G., Schröder, H.C., Arendes, J., Steffen, R., Zahn, R.K., Dose, K.: Alterations of activities of ribonucleases and polyadenylate polymerase in synchronized mouse L cells. Eur. J. Biochem. 76, 531-540 (1977a)
Müller, W.E.G., Schröder, H.C., Arendes, J., Zahn, R.K., Dose, K.: Filter paper disk techniques for assay of nucleotidase. Mol. Biol. Rep. 3, 331-337 (1977b)
Müller, W.E.G., Seibert, G., Beyer, R., Breter, H.J., Maidhof, A., Zahn, R.K.: Effect of cordycepin on nucleic acid metabolism in L5178y cells and on nucleic acid-synthesizing enzyme systems. Cancer Res. 37, 3824-3833 (1977c)
Müller, W.E.G., Arendes, J., Zahn, R.K., Schröder, H.C.: Control of enzymic hydrolysis of polyadenylate segment of messenger RNA: role of polyadenylate-associated proteins. Eur. J. Biochem. 86, 283-290 (1978a)
Müller, W.E.G., Falke, D., Zahn, R.K., Arendes, J.: Alterations of polyadenylate nuclease activities in herpes simplex virus-infected cells. Virology 87, 89-95 (1978b)
Müller, W.E.G., Zahn, R.K., Falke, D.: Variation of DNA and RNA polymerase activities in cells infected with herpes simplex virus type 1. Virology 84, 320-330 (1978c)
Müller, W.E.G., Zahn, R.K., Schröder, H.C., Arendes, J.: Age-dependent enzymatic poly(A) metabolism in quail oviduct. Gerontology 25, 61-68 (1979)
Müller, W.E.G., Schröder, H.C., Zahn, R.K., Dose, K.: Degradation of 2'-5'-linked oligoriboadenylates by 3'-exoribonuclease and 5'-nucleotidase. Hoppe-Seyler's Z. Physiol. Chem. 361, 469-472 (1980a)
Müller, W.E.G., Zahn, R.K., Arendes, J.: Age-dependent gene induction in quail oviduct. X. Alterations on the post-transcriptional level (enzymic aspect) Mech. Ageing Dev. 14, 39-48 (1980b)
Müller, W.E.G., Bernd, A., Schröder, H.C.: Modulation of poly(A)(+)mRNA-metabolizing and transporting-systems under special consideration of microtubule protein and actin. Mol. Cell. Biochem. 53/54, 197-220 (1983)
Müller, W.E.G., Agutter, P.S., Bernd, A., Bachmann, M., Schröder, H.C.: Role of post-transcriptional events in ageing: consequences for gene expression in eukaryotic cells. In: The 1984 Sandoz Lectures in Gerontology (eds. M. Bergener, M. Ermini, H.B. Staehelin). London-New York: Academic Press 1984
Murty, C.N., Verney, E., Sidransky, H.: The effect of tryptophan on nucleocytoplasmic translocation of RNA in rat liver. Biochim. Biophys. Acta 474, 117-128 (1977)
Murty, C.N., Verney, E., Sidransky, H.: Effect of tryptophan on nuclear envelope in rat liver: evidence for increased nuclear RNA release. In: The Nuclear Envelope and the Nuclear Matrix (ed. G.G. Maul), pp. 111-127. New York: Alan R. Liss 1982
Nabeshima, Y., Fujii-Kuriyama, Y., Muramatsu, M., Ogata, K.: Alternative transcription and two modes of splicing result in two myosin light chains from one gene. Nature (London) 308, 333-338 (1984)
Nakashima, K., Darzynkiewicz, E., Shatkin, A.S.: Proximity of mRNA 5'-region and 18S rRNA in eukaryotic initiation complexes. Nature (London) 286, 226-230 (1980)
Nevins, J.R.: Processing of late adenovirus nuclear RNA to mRNA. Kinetics of formation of intermediates and demonstration that all events are nuclear. J. Mol. Biol. 130, 493-506 (1979)

Nevins, J.R.: The pathway of eukaryotic mRNA formation. Annu. Rev. Biochem. 52, 441-466 (1983)
Nevins, J.R., Darnell, J.E.: Steps in the processing of Ad2 mRNA: poly(A)$^{+}$ nuclear sequences are conserved and poly(A) addition precedes splicing. Cell 15, 1477-1493 (1978)
Nichols, J.L., Welder, L.: A modified nucleotide in the poly(A) tract of maize RNA. Biochim. Biophys. Acta 652, 99-109 (1981)
Niessing, J., Sekeris, C.E.: Cleavage of high-molecular DNA-like RNA by a nuclease present in 30-S ribonucleoprotein particles of rat liver nuclei. Biochim. Biophys. Acta 209, 484-492 (1970)
Nilsen, T.W., Maroney, P.A., Robertson, H.D., Baglioni, C.: Heterogeneous nuclear RNA promotes synthesis of (2',5')oligoadenylate and is cleaved by the (2',5')oligoadenylate-activated endoribonuclease. Mol. Cell. Biol. 2, 154-160 (1982a)
Nilsen, T.W., Wood, D.L., Baglioni, C.: Presence of 2',5'-oligo(A) and of enzymes that synthesize, bind, and degrade 2',5'-oligo(A) in HeLa cell nuclei. J. Biol. Chem. 257, 1602-1605 (1982b)
Nudel, U., Soreq, H., Littauer, U.Z., Marbaix, G., Huez, G., Leclercq, M., Hubert, E., Chantrenne, H.: Globin mRNA species containing poly(A) segments of different lengths. Their functional stability in Xenopus oocytes. Eur. J. Biochem. 64, 115-121 (1976)
Ohtsuki, K., Groner, Y., Hurwitz, J.: Isolation and purification of double-stranded ribonuclease from calf thymus. J. Biol. Chem. 252, 483-491 (1977)
Ojala, D., Montoya, J., Attardi, G.: tRNA punctuation model of RNA processing in human mitochondria. Nature (London) 290, 470-474 (1981)
Ono, T., Cutler, R.G.: Age-dependent relaxation of gene repression: increase of endogenous murine leukemia virus-related and globin-related RNA in brain and liver of mice. Proc. Natl. Acad. Sci. USA 75, 4431-4435 (1978)
Padgett, R.A., Mount, S.M., Steitz, J.A., Sharp, P.A.: Splicing of messenger RNA precursors is inhibited by antisera to small nuclear ribonucleoprotein. Cell 35, 101-107 (1983)
Pederson, T.: Messenger RNA biosynthesis and nuclear structure. Am. Sci. 69, 76-84 (1981)
Peebles, C.L., Gegenheimer, P., Abelson, J.: Precise excision of intervening sequences from precursor tRNAs by a membrane-associated yeast endonuclease. Cell 32, 525-536 (1983)
Perry, R.P.: RNA processing comes to age. J. Cell Biol. 91, 28s-38s (1981)
Perry, R.P., Kelley, D.E., LaTorre, J.: Synthesis and turnover of nuclear and cytoplasmic polyadenylic acid in mouse L cells. J. Mol. Biol. 82, 315-331 (1974)
Perry, R.P., Schibler, U., Meyuhas, O.: The processing of messenger RNA and the determination of its relative abundance. In: From Gene to Protein: Information Transfer in Normal and Abnormal Cells (eds. T.R. Russell, K. Brew, H. Faber, J. Schultz), pp. 187-206. London-New York: Academic Press 1979
Phillips, C.R.: The regional distribution of poly(A) and total RNA concentrations during early Xenopus development. J. Exp. Zool. 223, 265-275 (1982)
Proudfoot, N.J., Brownlee, G.E.: 3' Non-coding region sequences in eukaryotic messenger RNA. Nature (London) 263, 211-214 (1976)
Prüsse, A., Louis, C., Alonso, A., Sekeris, C.E.: Isolation and characterization of hnRNA-snRNA-protein complexes from Morris hepatoma cells. Eur. J. Biochem. 128, 169-178 (1982)
Quintanilla, M., Renart, J.: Purification and characterization of a novel UpN-specific endoribonuclease VI from Artemia larvae. J. Biol. Chem. 257, 12594-12599 (1982)
Rech, J., Brunel, C., Jeanteur, P.: HnRNP from HeLa cells contain a ribonuclease active on double-stranded RNA. Biochem. Biophys. Res. Commun. 88, 422-427 (1979)

Reddy, E.S.P., Sitaram, N., Bhargava, P.M., Scheit, K.H.: A new pyrimidine-specific ribonuclease from bovine seminal plasma that is active on both single and double-stranded polyribonucleotides and that can distinguish between Mg^{2+}-containing and Mg^{2+}-depleted naturally occurring RNAs. J. Mol. Biol. 135, 525-544 (1979)
Reddy, R., Busch, H.: U snRNA's of nuclear snRNP's. In: The Cell Nucleus (ed. H. Busch), Vol. VIII, pp. 261-306. London-New York: Academic Press 1981
Reddy, R., Busch, H.: Small nuclear RNAs and RNA processing. Prog. Nucleic Acid Res. Mol. Biol. 30, 127-162 (1983)
Rhoads, R.E.: The cap structure of eukaryotic messenger RNA and its interaction with cap-binding protein. In: Progress in Molecular and Subcellular Biology (eds. F.E. Hahn, D.J. Kopecko, W.E.G. Müller), Vol. IX. Berlin-Heidelberg-New York: Springer 1985
Rinke, J., Steitz, J.A.: Precursor molecules of both human 5S ribosomal RNA and transfer RNAs are bound by a cellular protein reactive with anti-La lupus antibodies. Cell 29, 149-159 (1982)
Rinke, J., Blöcker, H., Appel, B., Lührmann, R.: The sequence at the 5' end of U1 snRNA proposed to base pair with an intron's 5' end is single-stranded in intact U1 snRNPs. 15th FEBS Meet. Brussels, Abstr. Book S-01, TU-078 (1983)
Robertson, H.D., Webster, R.E., Zinder, N.D.: Purification and properties of ribonuclease III from Escherichia coli. J. Biol. Chem. 243, 82-91 (1968)
Roeder, R.G., Rutter, W.J.: Specific nucleolar and nucleoplasmic RNA polymerases. Proc. Natl. Acad. Sci. USA 65, 675-682 (1970)
Rogers, J., Wall, R.: A mechanism for RNA splicing. Proc. Natl. Acad. Sci. USA 77, 1877-1879 (1980)
Rose, K.M., Jacob, S.T.: Nuclear poly(A) polymerase from rat liver and a hepatoma. Comparison of properties, molecular weights and amino acid compositions. Eur. J. Biochem. 67, 11-21 (1976)
Rose, K.M., Jacob, S.T.: Selective inhibitor of RNA polyadenylation by ara-ATP in vitro: a possible mechanism for antiviral action of ara-A. Biochem. Biophys. Res. Commun. 81, 1418-1424 (1978)
Rose, K.M., Jacob, S.T.: Phosphorylation of nuclear poly(A)polymerase. Comparison of liver and hepatoma enzymes. J. Biol. Chem. 254, 10256-10261 (1979)
Rose, K.M., Jacob, S.T.: Phosphorylation of nuclear poly(adenylic acid) polymerase by protein kinase: mechanism of enhanced poly (adenylic acid) synthesis. Biochemistry 19, 1472-1476 (1980)
Rose, K.M., Morris, H.P., Jacob, S.T.: Mitochondrial poly(A) polymerase from a poorly differentiated hepatoma: purification and characteristics. Biochemistry 14, 1025-1032 (1975)
Rose, K.M., Roe, F.J., Jacob, S.T.: Two functional states of poly (adenylic acid) polymerase in isolated nuclei. Biochim. Biophys. Acta 478, 180-191 (1977)
Rose, K.M., Jacob, S.T., Kumar, A.: Poly(A) polymerase and poly(A)-specific mRNA binding protein are antigenically related. Nature (London) 279, 260-262 (1979)
Rothman, F., Shatkin, A.J., Perry, R.P.: Sequences containing methylated nucleotides at the 5'termini of messenger RNA's. Cell 3, 197-199 (1974)
Roy, R.K., Sarkar, S., Guha, C., Munro, H.N.: RNP particles involved in release of in vitro synthesized poly(A)-containing RNA in isolated nuclei. In: The Cell Nucleus (ed. H. Busch), Vol. IX, pp. 289-308. London-New York: Academic Press 1981
Ruderman, J.V., Pardue, M.L.: Cell-free translation analysis of messenger RNA in echinoderm and amphibian early development. Dev. Biol. 60, 48-68 (1977)

Samarina, O.P., Krichevskaya, A.A.: Nuclear 30S RNP particles. In: The Cell Nucleus (ed. H. Busch), Vol. IX, pp. 1-48. London-New York: Academic Press 1981

Samarina, O.P., Lukanidin, F.M., Molnar, J., Georgiev, G.P.: Structural organization of nuclear complexes containing DNA-like RNA. J. Mol. Biol. 33, 251-263 (1968)

Sasavage, N.L., Smith, M., Gillam, S., Woychik, R.P., Rottman, F.M.: Variation of the polyadenylation site of bovine prolactin mRNA. Proc. Natl. Acad. Sci. USA 79, 223-227 (1982)

Schmid, H.-P., Schönfelder, M., Setyono, B., Köhler, K.: 76-kDa poly(A)-protein is involved in the formation of 48S initiation complexes. FEBS Lett. 157, 105-110 (1983)

Schmidt, A., Zilberstein, A., Shulman, L., Federman, P., Berissi, H., Revel, M.: Interferon action: Isolation of nuclease F, a translation inhibitor activated by interferon-induced (2'-5')oligo-isoadenylate. FEBS Lett. 25, 257-264 (1978)

Schmidt, A., Chernajovsky, Y., Shulman, L., Federman, P., Berissi, H., Revel, M.: An interferon-induced phosphodiesterase degrading (2'-5')oligoisoadenylate and the C-C-A terminus of tRNA. Proc. Natl. Acad. Sci. USA 76, 4788-4792 (1979)

Schröder, H.C., Dose, K., Zahn, R.K., Müller, W.E.G.: Isolation and characterization of the novel polyadenylate- and polyuridylate-degrading acid endoribonuclease V from calf thymus. J. Biol. Chem. 255, 5108-5112 (1980a)

Schröder, H.C., Zahn, R.K., Dose, K., Müller, W.E.G.: Purification and characterization of a poly(A)-specific exoribonuclease from calf thymus. J. Biol. Chem. 255, 4535-4538 (1980b)

Schröder, H.C., Schuster, D., Zahn, R.K., Müller, W.E.G.: A novel metabolic effect of the adenosine deaminase inhibitor coformycin, a potentiator of antiviral adenosine analogues. Antiviral Res. 1, 383-391 (1981)

Schröder, H.C., Bernd, A., Zahn, R.K., Müller, W.E.G.: Interaction of polyribosomal components and polyribonucleotides with microtubule proteins. Mol. Biol. Rep. 8, 233-237 (1982a)

Schröder, H.C., Zahn, R.K., Müller, W.E.G.: Role of actin and tubulin in the regulation of poly(A) polymerase-endoribonuclease IV complex from calf thymus. J. Biol. Chem. 257, 2305-2309 (1982b)

Schröder, H.C., Schenk, P., Baydoun, H., Wagner, K.G., Müller, W.E.G.: Occurrence of short-sized oligo(A) fragments during course of cell cycle and ageing. Arch. Gerontol. Geriatr. 2, 349-360 (1983)

Schröder, H.C., Bernd, A., Zahn, R.K., Müller, W.E.G.: Binding of polyribonucleotides and polydeoxyribonucleotides to bovine brain microtubule protein: age-dependent modulation via phosphorylation of high-molecular-weight microtubule-associated proteins and tau proteins. Mech. Ageing Dev. 24, 101-117 (1984a)

Schröder, H.C., Gosselin, G., Imbach, J.-L., Müller, W.E.G.: Influence of the xyloadenosine analogue of 2',5'-oligoriboadenylate on poly(A)-specific, 2',5'-oligoriboadenylate degrading 2',3'-exoribonuclease and further enzymes involved in poly(A)(+)mRNA metabolism. Mol. Biol. Rep., in press (1984b)

Schröder, H.C., Nitzgen, D.E., Bernd, A., Kurelec, B., Zahn, R.K., Gramzow, M., Müller, W.E.G.: Inhibition of nuclear envelope nucleoside triphosphatase regulated nucleocytoplasmic mRNA translocation by 9-β-D-arabinofuranosyladenine 5'-triphosphate. Cancer Res. 44, 3812-3819 (1984c)

Schwartz, H., Darnell, J.E.: The association of protein with the polyadenylic acid of HeLa cell messenger RNA: Evidence for a "transport" role of a 75,000 molecular weight polypeptide. J. Mol. Biol. 104, 833-851 (1976)

Seifert, H., Scheurlen, M., Northemann, W., Heinrich, P.C.: Low molecular weight RNAs as components of nuclear ribonucleoprotein particles containing heterogeneous nuclear RNA. Biochim. Biophys. Acta 564, 55-66 (1979)

Sekeris, C., Niessing, J.: Evidence for the existence of a structural RNA component in the nuclear ribonucleoprotein particles containing heterogeneous RNA. Biochem. Biophys. Res. Commun. 62, 642-650 (1975)

Sen, G.C.: Mechanism of interferon action. Prog. Nucleic Acid Res. Mol. Biol. 27, 105-156 (1982)

Sharp, P.A.: Speculations on RNA splicing. Cell 23, 643-646 (1981)

Shatkin, A.J.: Capping of eukaryotic mRNA's. Cell 9, 645-653 (1976)

Sheiness, D., Darnell, J.E.: Polyadenylic acid segment in mRNA becomes shorter with age. Nature (London) New Biol. 241, 265-268 (1973)

Sheiness, D., Puckett, L., Darnell, J.E.: Possible relationship of poly(A) shortening to mRNA turnover. Proc. Natl. Acad. Sci. USA 72, 1077-1081 (1975)

Sierakowska, H., Shugar, D.: Mammalian nucleolytic enzymes. Prog. Nucleic Acid Res. Mol. Biol. 20, 60-130 (1977)

Sippel, A.E., Stavrianopoulos, J.G., Schütz, G., Feigelson, P.: Translational properties of rabbit globin mRNA after specific removal of poly(A) with ribonuclease H. Proc. Natl. Acad. Sci. USA 71, 4635-4639 (1974)

Slattery, E., Ghosh, N., Samanta, H., Lengyel, P.: Interferon, double-stranded RNA, and RNA degradation: activation of an endonuclease by (2'-5')A_n. Proc. Natl. Acad. Sci. USA 76, 4778-4782 (1979)

Soreq, H., Nudel, U., Salomon, R., Revel, M., Littauer, U.Z.: In vitro translation of polyadenylic acid-free rabbit globin messenger RNA. J. Mol. Biol. 88, 233-245 (1974)

Sporn, M.B., Lazarus, H.M., Smith, J.M., Henderson, W.R.: Studies on nuclear exoribonucleases. III. Isolation and properties of the enzyme from normal and malignant tissues of the mouse. Biochemistry 8, 1698-1705 (1969)

Stark, B.C., Kole, R., Bowman, E.J., Altman, S.: Ribonuclease P: an enzyme with an essential RNA component. Proc. Natl. Acad. Sci. USA 75, 3717-3721 (1978)

Stefano, J.E.: Purified lupus antigen La recognizes an oligouridylate stretch common to the 3' termini of RNA polymerase III transcripts. Cell 36, 145-154 (1984)

Stein, J.L., Thrall, C.L., Park, W.D., Mans, R.J., Stein, G.S.: Hybridization analysis of histone messenger RNA: association with polyribosomes during the cell cycle. Science 189, 557-558 (1975)

Stoltzfus, C.M., Dane, R.W.: Accumulation of spliced avian retrovirus mRNA is inhibited in S-adenosylmethionine-depleted chicken embryo fibroblasts. J. Virol. 42, 918-931 (1982)

Therwath, A., Scherrer, K.: Post-transcriptional suppression of globin gene expression in cells transformed by avian erythroblastosis virus. Proc. Natl. Acad. Sci. USA 75, 3776-3780 (1978)

Tomcsányi, T., Komáromy, L., Tigyi, A.: Structural characterization of polysomal poly(A) protein particles in rat liver. Eur. J. Biochem. 114, 421-428 (1981)

Tomcsányi, T., Molnar, J., Tigyi, A.: Structural characterization of nuclear poly(A)-protein particles in rat liver. Eur. J. Biochem. 131, 283-288 (1983)

Tsiapalis, C.M., Dorson, J.W., De Sante, D.M., Bollum, F.J.: Terminal riboadenylate transferase: a polyadenylate polymerase from calf thymus gland. Biochem. Biophys. Res. Commun. 50, 737-743 (1973)

Tsiapalis, C.M., Dorson, J.W., Bollum, F.J.: Purification of terminal riboadenylate transferase from calf thymus gland. J. Biol. Chem. 250, 4486-4496 (1975)

Tsiapalis, C.M., Trangas, T., Gounaris, A.: Phosphorylation and activation of poly(A)-endoribonuclease from calf thymus gland. FEBS Lett. 140, 213-218 (1982)

Van Eekelen, C.A.G., van Venrooij, W.J.: HnRNA and its attachment to a nuclear protein matrix. J. Cell Biol. 88, 554-563 (1981)
Van Eekelen, C.A.G., Riemen, T., van Venrooij, W.J.: Specificity of the interaction of hnRNA and mRNA with proteins as revealed by in vivo crosslinking. FEBS Lett. 130, 223-226 (1981)
Van Venrooij, W.J., Sillekens, P.T.G., van Eekelen, C.A.G., Reinders, R.J.: On the association of mRNA with the cytoskeleton in uninfected and adenovirus-infected human KB cells. Exp. Cell Res. 135, 79-91 (1981)
Villarreal, L.P., White, R.T.: A splice junction deletion deficient in the transport of RNA does not polyadenylate nuclear RNA. Mol. Cell. Biol. 3, 1381-1388 (1983)
Vincent, A., Goldenberg, S., Scherrer, K.: Comparisons of proteins associated with duck-globin mRNA and its polyadenylated segment in polyribosomal and repressed free messenger ribonucleoprotein complexes. Eur. J. Biochem. 114, 179-193 (1981)
Wallace, J.C., Edmonds, M.: Polyadenylated nuclear RNA contains branches. Proc. Natl. Acad. Sci. USA 80, 950-954 (1983)
Wallace, J.C., Wood, W.M., Edmonds, M.: 5'-Terminal cap structures of oligo-(uridylic acid)-containing messenger ribonucleic acid from HeLa cells: comparison with other ribonucleic acid subpopulations. Biochemistry 20, 5364-5368 (1981)
Warnick, C.T., Lazarus, H.M.: The subcellular distribution of poly-A-degrading activity in mouse kidney. Can. J. Biochem. 55, 485-488 (1977)
Webb, T.E., Schumm, D.E., Palayoor, T.: Nucleocytoplasmic transport of mRNA. In: The Cell Nucleus (ed. H. Busch), Vol. IXB, pp. 199-248. London-New York: Academic Press 1981
Weber, J., Blanchard, J.M., Ginsberg, H., Darnell, J.E.: Order of polyadenylic acid addition and splicing events in early adenovirus mRNA formation. J. Virol. 33, 286-291 (1980)
Weinberg, R.A.: Nuclear RNA metabolism. Annu. Rev. Biochem. 42, 329-354 (1973)
White, J.P., Gardner, W.D., Hoch, S.O.: Identification of the immunogenically active components of the Sm and RNP antigens. Proc. Natl. Acad. Sci. USA 78, 626-680 (1981)
Williams, B.R.G., Kerr, I.M., Gilbert, C.S., White, C.N., Ball, L.A.: Synthesis and breakdown of pppA2'p5'A2'p5'A and transient inhibition of protein synthesis in extracts from interferon-treated and control cells. Eur. J. Biochem. 92, 455-462 (1978)
Winters, M.A., Edmonds, M.: A poly(A) polymerase from calf thymus. Purification and properties of the enzyme. J. Biol. Chem. 248, 4756-4762 (1973a)
Winters, M.A., Edmonds, M.: A poly(A) polymerase from calf thymus. Characterization of the reaction product and the primer requirement. J. Biol. Chem. 248, 4763-4768 (1973b)
Wold, B.J., Klein, W.H., Hough-Evans, B.R., Britten, R.J., Davidson, E.H.: Sea urchin embryo mRNA sequences expressed in the nuclear RNA of adult tissues. Cell 14, 941-950 (1978)
Wolin, S.L., Steitz, J.A.: Genes for two small cytoplasmic Ro RNAs are adjacent and appear to be single-copy in the human genome. Cell 32, 735-744 (1983)
Wood, W.M., Edmonds, M.: A method for isolation of oligo(uridylic acid)-containing messenger ribonucleic acid from HeLa cells. Biochemistry 20, 5359-5364 (1981)
Wreschner, D.H., McCauley, J.W., Skehel, J.J., Kerr, I.M.: Interferon action-sequence specificity of the ppp(A2'p)$_n$A-dependent ribonuclease. Nature (London) 289, 414-417 (1981)
Wreschner, D.H., Silverman, R.H., James, T.C., Gilbert, C.S., Kerr, I.M.: Affinity labelling and characterization of the ppp(A2'p)$_n$A-dependent endoribonuclease from different mammalian sources. Eur. J. Biochem. 124, 261-268 (1982)

Yang, V.W., Lerner, M.R., Steitz, J.A., Flint, S.J.: A small nuclear ribonucleoprotein is required for splicing of adenoviral early RNA sequences. Proc. Natl. Acad. Sci. USA 78, 1371-1375 (1981)

Zeevi, M., Nevins, J.R., Darnell, J.E.: Newly formed mRNA lacking polyadenylic acid enters the cytoplasm and the polyribosomes but has a shorter half-life in the absence of polyadenylic acid. Mol. Cell. Biol. 2, 517-525 (1982)

Zeller, R., Nyffenegger, T., DeRobertis, E.M.: Nucleocytoplasmic distribution of snRNPs and stockpiled snRNA-binding proteins during oogenesis and early development in Xenopus laevis. Cell 32, 425-434 (1983)

Zieve, G., Penman, S.: Small RNA species of the HeLa cell: metabolism and subcellular localization. Cell 8, 19-31 (1976)

The Cap Structure of Eukaryotic Messenger RNA and its Interaction with Cap-binding Protein

R. E. Rhoads

A. Introduction

This article is a review of what is known about the 7-methylguanosine-containing "cap" structure of eukaryotic messenger RNA and its participation in the initiation of protein synthesis. Particular attention will be paid to the protein which is thought to mediate the entry of mRNA into the cycle of protein synthesis by recognizing the cap structure, termed cap-binding protein (CBP). Previous review articles dealing with this or related topics include those of Shatkin (1976), Banerjee (1980), Ehrenfeld (1982), Penman (1982) and Nielsen et al. (1983). The primary focus of this review will be translational events. Topics which will not be treated include biosynthesis of caps (Rottman 1978; Banerjee 1980), the role of the cap structure in stabilization of mRNA against degradation (Shimotohno et al. 1977; Furuichi et al. 1977), and the CBP which is involved in initiation of transcription of influenza virus (Blaas et al. 1982).

Recent advances in this area have created a great deal of interest in the molecular components and events involved in recognition of the 5'-terminus of mRNA. One of these advances has been the discovery that protein synthesis in eukaryotic and prokaryotic organisms differs fundamentally in the manner by which initiation codons are selected: whereas the prokaryotic ribosome can initiate at internal sites on an mRNA, its eukaryotic counterpart begins at the extreme 5'-terminus of the mRNA and scans the sequence for the first AUG codon (reviewed by Kozak 1978). Thus, binding to the cap is the first involvement of mRNA in the initiation process and, as such, represents the committed step in protein synthesis. Therefore, this step is potentially an ideal site for regulatory events, either of the overall rate of polypeptide initiation or of the discrimination of one mRNA over another. The question of competition for translation between different mRNA's has long been a subject of interest, but molecular mechanisms have not yet been elucidated (see refs. in Walden and Thach 1982). Another focus of interest is the observation that the cap recognition step appears to be a target for inactivation during viral infection. Poliovirus, whose genome consists of an uncapped single-stranded RNA, renders the initiation machinery incapable of recognizing capped cellular mRNA's yet able to translate the poliovirus RNA. Still another intriguing fact to this problem is the cytoskeleton; evidence has been presented that the CBP is affixed to the intermediate filament network and thereby links all mRNA of the cell to this structural framework.

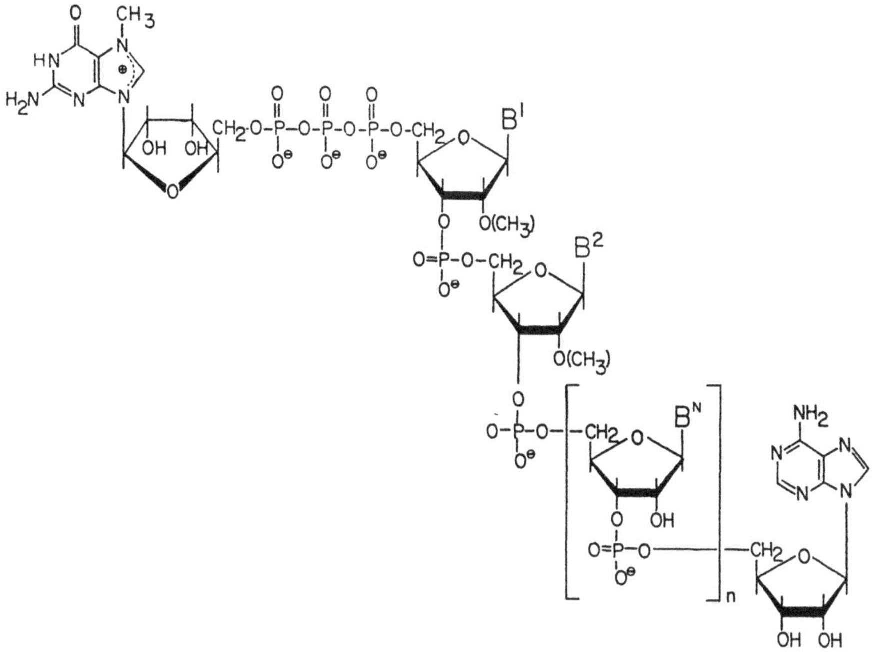

Fig. 1. Structure of eukaryotic mRNA showing the 7-methylguanosine-containing cap structure

The following review will first cover the unique chemical and conformational properties of 7-methylguanosine and the cap structure. This will be followed by a discussion of the manner in which the presence of a cap structure on mRNA enhances the rate of initiation of translation. Next will be a discussion of how CBP may be detected and purified and how it is thought to interact with other initiation factors. Finally, some special topics will be considered: antibodies directed against CBP or associate elements, the effect of viral infection on cap recognition, and the possible localization of mRNA and CBP on the cytoskeletal network.

B. 5'-Terminal Cap Structure of Messenger RNA

I. Structure and Occurrence in Nature

Eukaryotic messenger RNA contains a unique structural feature at its 5'-terminus consisting of 7-methylguanosine in a 5'-to-5' linkage via a triphosphate bridge to the first coded base of the mRNA (Rottman et al. 1974; Furuichi and Miura 1975; J.M. Adams and Cory 1975). This structure has been termed a "cap" (Rottman et al. 1974) and is pictured in Figure 1. The methylation in the 7-position of guanine always occurs. Methylations of the 2'-O-positions of the first (B^1) or second (B^2) nucleoside residues are variable, even for a single type of mRNA. The most abundant form of a given cellular mRNA, e.g., hen ovalbumin mRNA (Malek

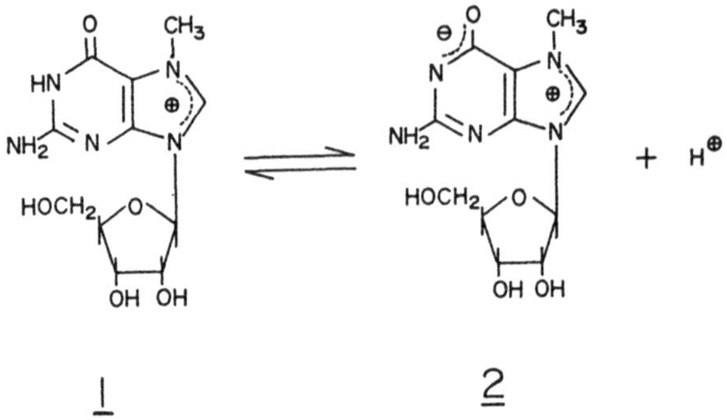

Fig. 2. Ionization of 7-methylguanosine. 1 *keto* form, 2 *enolate* form

et al. (1981), is generally methylated in both positions. The first nucleoside (B^1) is coded for by the gene, but purines are favored over pyrimidines, and 6-methyladenosine is the most common nucleoside found.

The distribution of capped mRNA's in nature has been reviewed (Shatkin 1976; Banerjee 1980). Briefly, all eukaryotic cellular mRNA's examined to date are capped. In addition, the small nuclear RNA's of the U-class contain a very similar structure, in which 2,2,7-trimethylguanosine replaces 7-methylguanosine (Shibata et al. 1975). Most viruses also contain capped mRNA, but there are notable exceptions: among the animal viruses, picornaviruses (poliovirus, EMC virus, and mengovirus), and among the plant viruses, tobacco necrosis virus, its satellite, STNV (satellite tobacco necrosis virus), and cowpea mosaic virus (see references in Banerjee 1980). Recently, other plant viruses have been added to the list: the nepoviruses (Mayo et al. 1979) and the potyviruses (Hellmann et al. 1980; Hari 1981). Those RNA's which lack a cap are usually terminated with a covalently linked protein.

II. Chemical Properties of Caps

Understanding the biological interactions between the 5'-terminus of mRNA and cellular components requires an understanding of the unique chemical features of the cap moiety itself. Many of the approaches that have been taken to elucidate the role and action of caps have taken advantage of chemical differences between the cap moiety and normal nucleotide residues in mRNA. The following is a partial list of some of the more interesting chemical properties.

1. Ionization of the N-1 Proton

Alkylation of the 7-position of guanosine introduces a positive charge into the ring system (Fig. 2). This has the effect of lowering the pK_a for loss of the N-1 proton from approx. 9.4 in guanosine (Shapiro 1968) to 6.7 in 7-methylguanosine (Hendler et al. 1970). Thus, whereas guanosine is uncharged at neutral pH, 7-

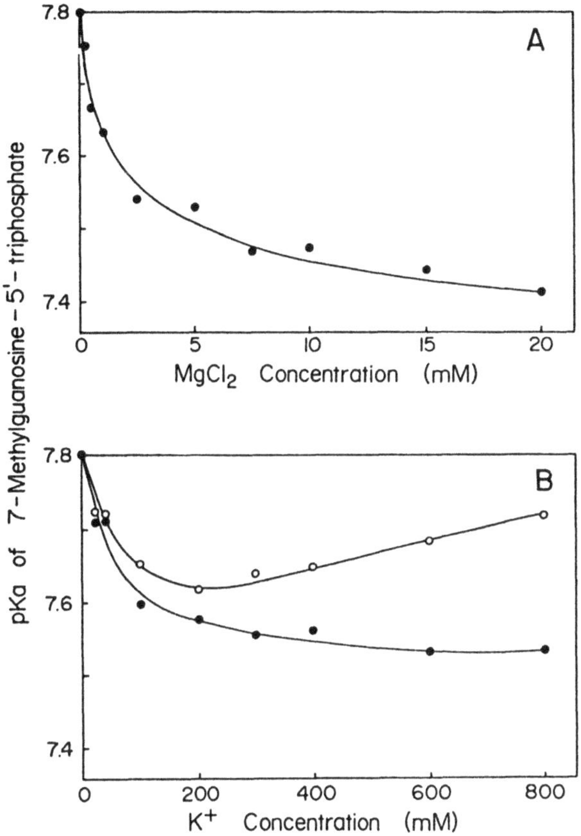

Fig. 3. Effect of salts on the pK_a of the N-1 position of m^7GTP. The pK_a may be determined from a single measurement of the $A_{258}:A_{282}$ ratio (R) at a given pH, using the equation

$$pK_a = pH - \log\left(\frac{Rc_1 - c_2}{c_3 - c_4 R}\right)$$

where c_1 is the ε_M^{282} of compound 1 (Fig. 2) = 7.34×10^3, c_2 is the ε_M^{258} of compound 1 = 11.4×10^3, c_3 is the ε_M^{258} of compound 2 = 6.05×10^3, and c_4 is the ε_M^{282} of compound 2 = 8.78×10^3. All measurements were made in 10 mM sodium HEPES buffer, pH 7.5. (A) variation of pK_a with $MgCl_2$ concentration. (B) variation of pK_a with KCL (●-●) and K(OAc) (o-o). This experiment was performed in collaboration with G. Hellmann, University of Kentucky, and P. Remy, I.B.M.C. Strasbourg, France

methylguanosine is a mixture of cation (1) and zwitterion (2). The introduction of phosphate moieties at the 5'-position causes the pK_a to again rise. Thus, the pK_a's of m^7GMP, m^7GDP, and m^7GTP were reported to be, respectively, 7.1, 7.2, and 7.5 (Hendler et al. 1970). This effect may be rationalized as a partial cancellation of the positive charge on N-7 by the negatively charged phosphate moieties. In fact, direct evidence for an electrostatic interaction between N-7 and the phosphate of m^7GMP has been obtained (see below).

Certain factors affect the value of the pK_a. Nishimura et al. (1980) reported the pK_a of m^7GTP (presumably in 0.1 M NaCl) to be 7.6 at 11°C and 7.3 at 30°C. We report here that salts also affect the pK_a. As shown in Figure 3, the pK_a of m^7GTP was 7.8 (at 25°C) in the absence of salt, but decreased to 7.4 with increasing $MgCl_2$ concentration. KCl, at higher concentrations, similarly caused the pK_a to decrease from 7.8 to 7.5. K(OAc) unexpectedly caused the pK_a to first decrease to approximately 7.6 and then rise again at higher concentrations. One way to view this phenomenon is that the negative charge of the phosphates is being shielded at higher ionic strength and cannot exert the same neutralizing effect on the positive charge at N-7.

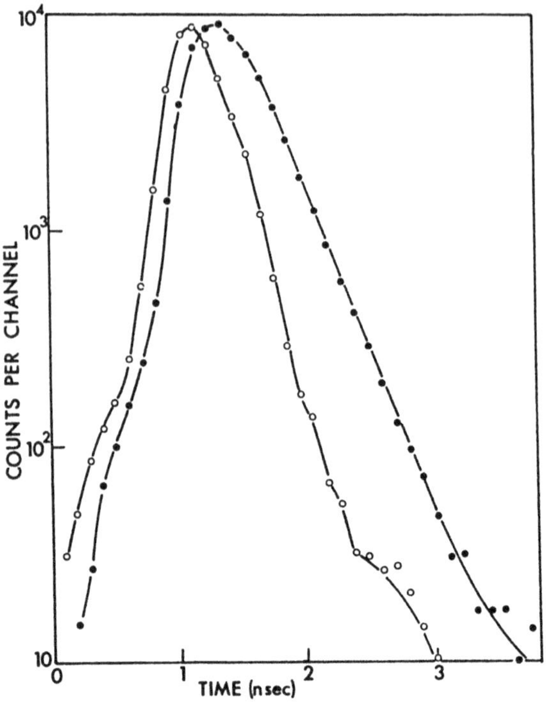

Fig. 4. Fluorescence lifetime of m^7GTP in aqueous solution. A 500 µM solution of m^7GTP in 10 mM MES buffer, pH 6.01, was irradiated at 292 nm with an argon laser and fluorescence was measured at 370 nm. •, fluorescence of m^7GTP. o, flash alone. This experiment as well as those in Figs. 5 and 6 were performed in collaboration with P. Remy, I.B.M.C., Strasbourg, France

2. Fluorescence

A unique property of m^7G among the bases in mRNA is the fact that it is fluorescent. Whereas the normal bases have very low quantum yields in neutral aqueous solutions at room temperature, ranging from 2×10^{-6} to 3×10^{-5}, the value for m^7G is 1.5×10^{-2} (Nishimura et al. 1977). The corrected excitation spectrum is similar to the absorption spectrum, with maxima at 258 and 282 nm. Both excitation and emission spectra are pH dependent; the 258-nm peak is predominant for the *keto* form (1) while the 282-nm peak is predominant for the *enolate* form (2).

Nishimura et al. (1980) have estimated the fluorescence lifetime of m^7Guo to be 250 ps based on the quantum yield. We have measured the lifetime directly and report here that the fluorescence decays monoexponentially with half-life of 200 ps in aqueous solution (Fig. 4). In dioxane, however, the quantum yield is nearly 20-fold higher. The fluorescence decay in this solution is biexponential, with lifetimes of 0.76 ns and 4.3 ns (Fig. 5). The relative contribution of the two lifetimes is dependent on pH: in acidic pH, the long lifetime accounts for 3% of the decay whereas in basic pH, it accounts for 20% - 40% (Fig. 6).

3. Conformation

Hickey et al. (1977) reported from NMR studies that, whereas there is an equal mixture of the ^{2}E and ^{3}E conformers of ribose

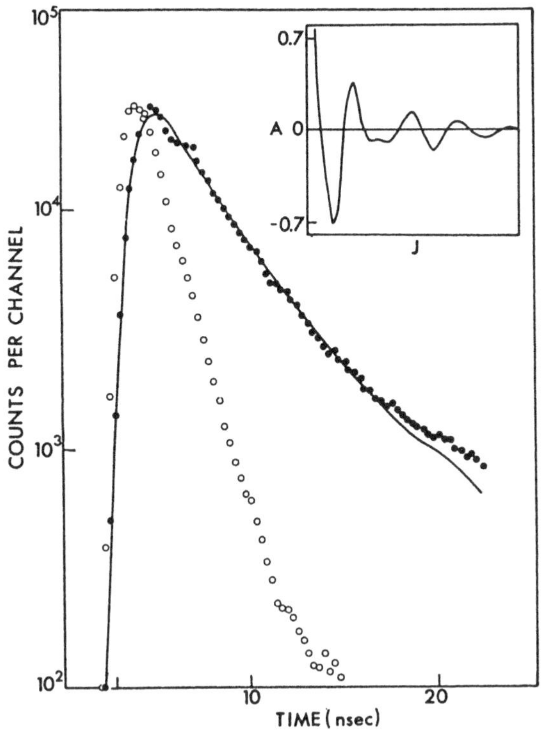

Fig. 5. Fluorescence lifetime of m^7GTP in dioxane. A 50 μM solution was excited at 310 nm and fluorescence measured above 370 nm with a single source (pulse width 2.5 ns). Computer analysis of the data revealed that a single exponential decay law did not fit. The curve shown is a two-exponential least-squares fit, calculated by computer as a convolution product with the flash. The two half-lives are 0.76 ns. contributing 75%, and 4.34 ns, contributing 25%. ●, fluorescence of m^7GTP. o, flash alone. *Inset*, autocorrelation function *(A)* of data

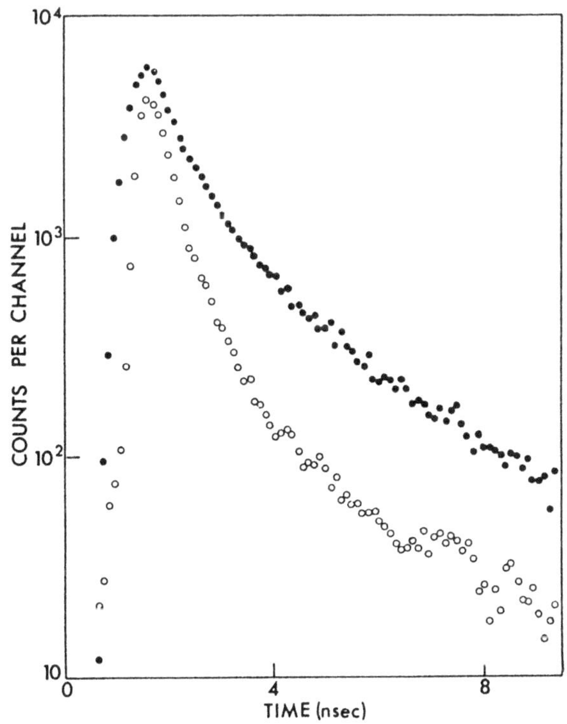

Fig. 6. Fluorescence lifetime of 7-methylguanosine in dioxane as a function of pH. A 50 μM solution of 7-methylguanosine in the presence of either 1% triethylamine (●) or 1% acetic acid (o) was excited at 292 nm with an argon laser (pulse width 10 ps) and fluorescence detected at 330 nm. The contribution of the long lifetime was graphically estimated as 20% - 40% in the presence of triethylamine but only 3% in the presence of acetic acid

in 7-methylguanosine, the effect of adding the 5'-phosphate is to stabilize the ^{3}E conformer so that it represents 2/3 of the total. They interpreted this as evidence for proximity of the phosphate group and N-7. Further NMR studies with the dinucleotide $m^{7}GpppAm$ provided evidence for base-stacking of the two nucleosides (Kim and Sarma 1977). This conclusion was also reached from circular dichroism studies on $m^{7}GpppA$ and $m^{7}GpppAm$ by Hattori et al. (1978). These authors also reported that the degree of stacking is increased both in acidic pH and also by the presence of the 2'-O-methyl group on adenosine. Circular dichroism spectra were also highly dependent on salt and temperature.

Nishimura et al. (1980) were able to study the base-stacking phenomenon using the fluorescence of $m^{7}G$, since fluorescence was found to be quenched in the stacked conformation. Factors which were found to favor stacking were decreased temperature, decreased pH, the presence of a purine nucleoside (N) in the structure $m^{7}GpppN$, and 2'-O-methylation of residue N. The transition from stacked to unstacked conformation as a function of pH coincided with the conversion of the *keto* form of $m^{7}G$ to the *enolate* form (see Fig. 2).

4. Reactions of the Ribose Moiety

Since the $m^{7}Guo$ is linked *via* the 5'-OH of ribose to mRNA, the 2'- and 3'-OH groups are free. This permits oxidation by periodate to the 2',3'-dialdehyde. Such ribose-ring oxidized mRNA's have been useful in several ways for studying the structure and function of caps. The aldehydes may be reduced with $NaB[^{3}H]_4$ to introduce radioactivity into the mRNA at the cap (Furuichi et al. 1975; Muthukrishnan et al. 1975; Lockard and Rajbhandary 1976; Breter et al. 1979). Also, the cap may be removed by β-elimination by treatment of oxidized mRNA with aniline (Furuichi et al. 1975; Wei and Moss 1975) or cyclohexylamine (Lockard and Rajbhandary 1976). A third way that oxidized mRNA may be used is to cross-link with proteins which have an affinity for the cap (see Sect. D.I.1).

5. Reactions of 7-methylguanine

Alkylation at N-7 of guanosine, in addition to lowering the pK_a of the N-1 proton (Sect. B.II.1), also renders the N-glycosidic bound unstable to hydrolysis in acid and the imidazole ring, to hydrolysis in alkali (reviewed by Hendler et al. 1970). The imidazolium cation is also susceptible to reduction with $NaBH_4$ (Hecht et al. 1976), and, in the 7,8-dihydro form, may be photodegraded at wavelengths which are not absorbed by other bases (Pochon et al. 1970). 7-methylguanosine also reacts with methylglyoxal approx. two orders of magnitude faster than does guanosine (Kozarich and Deegan 1979).

C. The Dependence of Translation on the Cap Structure

Determining the role of the cap structure in translation necessitates a study of very complex systems. There are literally hundreds of polypeptides and RNA's which act coordinately in the synthesis of proteins, and it is difficult to simplify the system appreciably in studying the overall properties of mRNA translation without losing essential elements. Thus, correctly interpreting the action of a single m^7Guo residue in this diversity of reactions is not straightforward. Two broad approaches have been used. The first is the comparison of mRNA's lacking a cap with normal capped mRNA's. In this context, a "cap" is considered to be a complete m^7Gppp-moiety. mRNA's which contain GpppN... termini are functionally uncapped, since the 7-methylation is required for cap function (see Sect. C.II). The second approach is the use of cap analogues as inhibitors of cell-free translation. In this section, these experiments will be discussed together with factors which affect the relative importance of caps in translational systems. The literature on this area has been extensively reviewed by Banerjee (1980). To avoid as much repetition as possible, the salient points will be mentioned with a minimum of literature references. For additional references, the reader is directed to the earlier review.

I. Studies on mRNA Lacking a Cap

1. Synthesis of mRNA's Lacking a Cap

Viral systems have proved invalualbe for the production of modified RNA's. Both et al. (1975a) used reovirus and VSV to synthesize mRNA's, using endogenous enzymes. mRNA synthesized in the presence of the methyl donor S-adenosylmethionine was roughly fivefold more active in a wheat germ cell-free translational system than that synthesized in its absence. Both et al. (1975b) subsequently showed that reovirus mRNA's synthesized in the absence of S-adenosylmethionine were terminated predominantly with ppG while those synthesized in its presence were terminated with m^7GpppGm. The level of m^7GpppGm caps could be even further suppressed by including S-adenosylhomocystine. These authors went on to show that methylated mRNA's were readily incorporated into polysomes whereas their unmethylated counterparts were not. The block appeared to be at or before the binding of mRNA to 43S initiation complexes (see Sect. E.I.I. for nomenclature used with initiation complexes).

Another way that the behavior of uncapped mRNA has been studied by synthesis of artificial mRNA's using the reovirus-derived oligonucleotides GpppGpC or ppGpC as primers for polynucleotide phosphorylase (Both et al. 1976). Whereas 58% of random polymers composed of A and U and terminated with m^7GpppGmpC could be bound to wheat germ ribosomes, those terminated with GpppCpC, ppGpC and p could be bound, respectively, only 17%, 13%, and 24%. Similar results were obtained with other random polymers.

2. Removal of Caps from mRNA

Preparing uncapped mRNA by omitting a methyl donor has been possible only in certain viral systems. An alternative is to specifically remove the cap from intact mRNA. Two general approaches to this problem have been taken: chemical and enzymatic.

As mentioned above (Sect. B.II.4), the 7-methylguanosine moiety in caps can be removed by β-elimination. Muthukrishnan et al. (1975) treated reovirus mRNA, in which 75% of the termini were m^7GpppGm and 25% were ppG, with periodate and aniline. This resulted in 87% removal of m^7Guo. The translational ability of the mRNA in a wheat germ system was decreased by 85%. This suggests a very high, if not absolute, dependence on the cap. What is not clear, however, is whether periodate and aniline caused degradation of the mRNA, compounding the inhibition. The authors state, however, that there was no change in sedimentation properties of the mRNA. Rose and Lodish (1976) also used β-elimination to remove the cap of VSV mRNA, but under somewhat milder conditions. They reported that some degradation of the mRNA was apparent. Bacteriophage RNA was used as a control for nonspecific alteration of RNA and was shown to lose 20% - 50% of its translational activity by the treatment. VSV mRNA treated with periodate alone or periodate plus aniline retained 33% and 25%, respectively, of its translational activity in a reticulocyte cell-free system. Less striking reductions in activity were obtained for polysome and 80S initiation complex formation. The authors concluded that mRNA, under the conditions of their experiments, exhibited only a two- to fourfold dependency on the cap, considerably less than observed when uncapped mRNAs were synthesized de novo (Sect. C.I.1). Chu et al. (1978) used an even gentler procedure for chemical decapping, substituting cyclohexylamine for aniline, and observed a six- to eightfold decrease in activity of globin and ovalbumin mRNA.

Caps may also be removed from mRNA enzymatically. Zan-Kowalczewska et al. (1977) treated several mRNA's with potato nucleotide pyrophosphatase. Ribosome binding in the case of reovirus mRNA was reduced more than tenfold. Translational activity in a wheat germ system was reduced tenfold in the case of rabbit globin mRNA and approx. fivefold for reovirus and *Artemia salina* mRNA. Another enzyme which has been used for this purpose is tobacco acid pyrophosphatase (Efstratiadis et al. 1977). Translational activity of mRNA decapped in this way was five- to tenfold reduced (Shimotohno et al. 1977; Chu et al. 1978). A report appeared that bacteriophage T_4-induced polynucleotide kinase also could cleave the cap structure (Abraham and Lillehaug 1976) but this has not yet been confirmed.

II. Cap Analogues as Inhibitors of Protein Synthesis

Another major tool in elucidating the role of mRNA caps has been the use of cap analogues. Compounds such as m^7GMP, m^7GDP, m^7GTP, and m^7GpppA have been shown to be potent inhibitors of protein synthesis (Hickey et al. 1976; Canaani et al. 1976). Evidence for specificity comes from the finding that capped mRNA's, such as those of SV-40, globin, ovalbumin, etc., are more sensitive to inhibition than naturally uncapped mRNA's, such as those of EMC (encephalomyocarditis) virus, STNV, mengo virus, etc. (Canaani et al. 1976; Hickey et al. 1976; Weber et al. 1976; Sharma et al. 1976). Inhibition was shown to be at the level of initiation, since cap analogues reduced the binding of mRNA to 80S ribosomes, provided the elongation inhibitor sparsomycin was also present (Hickey et al. 1976; Canaani et al. 1976). Further evidence for a specific effect on initiation is the fact that cap analogues do not affect formation of the 43S initiation complex (see Sect. E.I.1), but do prevent the conversion of the 43S complex to the 80S complex, a step which involves the joining of mRNA to the ribosome (Roman et al. 1976). No effect of cap analogues on polypeptide chain elongation, however, can be detected (Suzuki 1976). Furthermore, the inhibition by cap analogues appears to be competitive: increasing the concentration of mRNA overcomes inhibition (Willems et al. 1979; Chu and Rhoads 1980). All of these observations taken together mean that mRNA binds to some factor via the cap during formation of 48S initiation complexes; cap analogues prevent this binding competitively by occupying the same site.

In spite of this straightforward interpretation, there are some unexpected effects of cap analogues. Seal et al. (1978) reported that cap analogues inhibit translation of the *uncapped* RNA of STNV (49% inhibition at 500 μM m^7GMP), though not as severely as that of capped mRNA. Chu et al. (1978), on the other hand, showed that STNV RNA translation was stimulated by cap analogue (30% stimulation at 100 μM m^7GTP). Hickey et al. (1976) observed both effects: m^7GMP stimulated STNV RNA translation up to 250 μM and thereafter inhibited it. Cap analogues also produce a small but reproducible stimulation of (capped) ovalbumin mRNA translation at low concentrations (10 μM m^7GTP; Chu and Rhoads 1980). Also, Willems et al. (1979) reported that the uncapped RNA of mengovirus is inhibited by cap analogues, but only in the wheat germ cell-free system; translation in a reticulocyte system was not affected.

These results demonstrate that one must use caution in the interpretation of results with cap analogues. The ability of an mRNA's translation to be inhibited by cap analogues should not be taken as absolute proof that the mRNA is capped. Some factors which must be taken into consideration are discussed below (Sect. C.V).

III. The Effects of in vitro Addition of Caps to mRNA

A third approach to demonstrate the effects of the cap structure on translation is to enzymatically add the cap. Muthukrishnan et al. (1978) β-eliminated VSV (vesicular stomatitis virus) mRNA and observed at least a tenfold decrease in either translation or ribosome binding. They then used vaccinia capping enzymes to resynthesize the cap and were able to restore activity to within 90% of the original. Smith and Clark (1979) used this technique on the naturally uncapped RNA of STNV. In contrast to the above results, the rate of initiation complex formation with wheat germ ribosomes was identical for the capped and uncapped samples. This indicates that additional features of mRNA structure are important for initiation of translation, a subject that is discussed below (Sect. E.III.3). Rosenberg and Paterson (1979) took this approach one step further and capped prokaryotic mRNA. They found that this improved the translational efficiency of the mRNA in a wheat germ system and also preferentially directed translation to the first (5'-proximal) coding region.

IV. Structural Features of the Cap which are Important for Function

The three previous sections have indicated ways in which the participation of caps in translation may be assessed. It is now possible to discuss the effect of modifying the cap structure on its activity in vitro.

1. Phosphate Groups

At least one phosphate group is required at the 5'-position of 7-methylguanosine for cap activity; neither 7-methylguanosine itself nor 7-methylguanosine 2 (3)'-monophosphate were active as a cap analogue inhibitors (Hickey et al. 1976; Canaani et al. 1976). Adding the β and γ phosphates produces a more effective inhibitor, but the presence of the second nucleoside residue ($m^7G^{5'}ppp^{5'}N$) does not further improve the potency of the inhibitor (Hickey et al. 1977; Suzuki 1977; Chu and Rhoads 1978). The most effective cap analogue inhibitor found to date is the symmetrical tetraphosphate $m^7Gppppm^7G$ (Sasavage et al. 1978). The phosphate moiety of m^7GMP may be esterified with methanol, and surprisingly, the methyl ester is considerably less active than m^7GMP as a cap analogue inhibitor (Darzynkiewicz et al. 1981). This compound, nonetheless, retains the rigid ribose conformation of m^7GMP, as shown by NMR studies (see Sect. B.II.3). Thus, the overall negative charge, which is reduced by one in the inactive methyl ester, may be important in cap recognition aside from its putative role in conferring a rigid ribose conformation.

2. Ribose Ring

This portion of the cap does not appear to be directly involved in function since (a) 2'-deoxy-7-methylguanosine 5'-diphosphate is an active cap analogue (B.L. Adams et al. 1978) and (b) derivatization of the 2',3'-positions with levulinic acid yields an affinity medium which can bind CBP (see Sect. D.II.2). However, as mentioned above, mRNA in which the 2'- and 3'-hydroxyls had been oxidized to aldehydes with periodate was less active than untreated mRNA (Rose and Lodish 1976). Whether this was due to nonspecific mRNA degradation or modification of the cap was not determined. In a later study by Muthukrishnan et al. (1976), however, periodate-oxidized mRNA was shown to be nearly as active as the control. Some of the discrepancy between this result and that of Rose and Lodish (1976) was due to the source of the cell-free system (wheat germ *vs.* reticulocyte) and is discussed below (Sect. C.V.6).

Although the ribose moiety does not appear to be as directly involved as other parts of the cap, an effect of 2'-O-methylation has been noted. Muthukrishnan et al. (1976) compared synthetic polynucleotides capped with either m^7GpppG or m^7GpppGm for binding to ribosomes. With each type of random polynucleotide tested, the methylated form bound considerably better. In a later study, however, Muthukrishnan et al. (1978) found that natural mRNA's (VSV) synthesized with limiting methyl donor did not exhibit marked differences; binding of 2'-O-methylated and unmethylated mRNA's to ribosomes was nearly the same.

3. Guanine Moiety

The necessity of a methyl group in the 7-position was shown in the initial demonstration that cap analogues were inhibitors of protein synthesis since GMP was inactive (Hickey et al. 1976; Canaani et al. 1976). B.L. Adams (1978) subsequently showed that it was not the methyl group *per se* that was important but rather the presence of a positive charge on the imidazolium ring, since 7-ethyl- and 7-benzylguanosine 5'-diphosphate were as active as m^7GDP in inhibiting protein synthesis. This finding is also consistent with the earlier observation that polynucleotides terminated with a normal cap bind readily to ribosome, but those terminated with an alkali-treated cap, in which the imidazole ring is opened with consequent loss of the positive charge, are inactive (Both et al. 1976). Also mRNA synthesized with a 7-ethylguanosine cap is fully active (Furuichi et al. 1979). Other modifications of the imidazole ring made by B.L. Adams et al. (1978) included methylation of the 8-carbon, which decreased activity slightly, and hydride addition to the 8-carbon, which abolished the positive charge as well as cap analogue inhibitory activity, in agreement with the above-stated principle that the charge is necessary.

Modifications of the 6-membered ring of guanine included methylation of the 2-amino group, which did not alter activity, removal of the 2-amino group (m^7IDP), which resulted in a substantial reduction of activity, and substitution of the 2-amino group with oxygen (m^7XDP), which abolished activity. Two particularly interesting analogues, in light of the ionization of the N-1 position and its possible implications for cap recognition (Sect. C.V.3), were $m_2^{1,7}$-GDP and 6-Cl-m^7 (Fig. 7), both of which were inactive. The former compound is unable to lose a proton from N-1; the latter is unable to bear a negative charge on the 6-oxygen. Thus, neither analogue can exist in the *enolate* form (Fig. 2). A final modification of the 6-membered ring which renders the cap inactive, described by Kozarich and Deegan (1979), is addition of methylglyoxal. This forms a tricyclic structure by bridging between the N-1 and 2-amino group. Like the $m_2^{1,7}$-GTP compound discussed above, the tricyclic adduct is unable to ionize at the N-1 position.

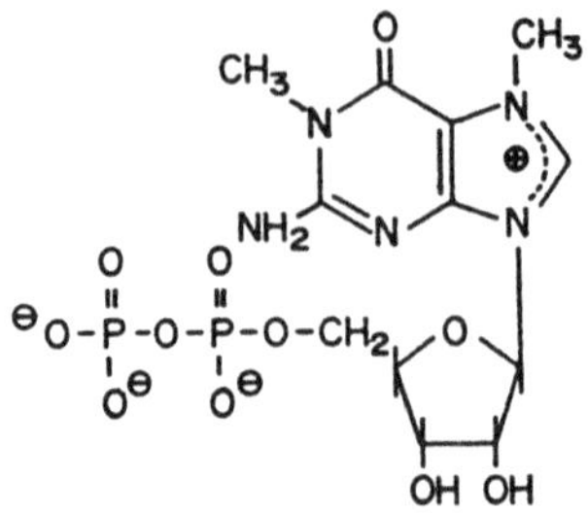

3

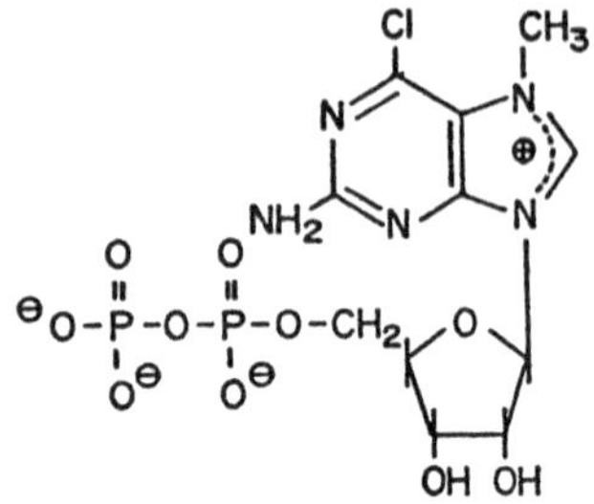

4

Fig. 7. Cap analogues synthesized by B.L. Adams et al. (1978). 3 1,7-dimethylguanosine 5'-diphosphate. 4 6-chloro-7-methylguanosine 5'-diphosphate

To summarize the results discussed in this section, there appears to be either a specific conformation or a specific charge distribution for the cap moiety to participate in translation. The results with the inactive m^{7}GMP methyl ester, in which the charge is reduced but the conformation remains unaltered, indicate that if conformation is important, it is nevertheless overridden by alteration of charge. If the *enolate* form of the cap is indeed the one recognized, then there is a negative-positive-negative charge distribution centering, respectively, around (a) the N-1, 0-6 locus, (b) the N-7, N-9 locus, and (c) the α-phosphate. In addition to these features, the 2-amino group greatly enhances cap activity.

V. Factors Which Influences the Translational Recognition of Caps

From the results presented in Sections C.I - III, it is clear that the presence of a cap on mRNA enhances the rate of initiation, but it is not an obligatory requirement for translation. In the course of demonstrating this dependancy, a large number of variables were discovered which influence the quantitative aspects of this enhancement. While a knowledge of these variables is useful for interpreting cell-free translational data, the greater utility lies in the insights they may give into the molecular details of the cap-recognition process. This section will deal with the best documenta of these variables.

1. Ionic Strength

Protein-nucleic acid interactions generally involve ionic forces. To the extent that they do, these interactions are weakened by increasing the ionic strength of the medium, since charges on the protein or nucleic acid are shielded by salt ions. Therefore, it was unexpected to find that caps are recognized *more* as the ionic strength is increased (Weber et al. 1977; Kemper and Stolarsky 1977; Chu and Rhoads 1978). Cap analogues do not inhibit translation if the K(OAc) concentration of the cell-free system is below about 50 mM but are maximally effective inhibitors at 200 mM. Also, decapped mRNA has the same translational activity as capped mRNA at low ionic strength but is considerably less active at high ionic strength. These results may be interpreted to mean that there is a factor through which the stimulatory effect of the cap is mediated, and that the binding of the cap to this factor is greater at higher ionic strength than at lower. The corollary to this is that a cap-independent pathway of initiation must exist, i.e., the cap plays a facilitative, but not obligatory, role. At low ionic strength the cap is not recognized by the factor, but initiation occurs nonetheless, albeit more slowly.

2. Temperature

Weber et al. (1978) examined the effect of the temperature of incubation of the reticulocyte cell-free system on cap recognition. Cap analogues became progressively poorer inhibitor as the temperature was raised. As wheat germ systems are generally incubated at lower temperatures than those of reticulocyte, this may partly account for some of the observed differences in cap recognition between the two systems (see Sect. C.V.6).

3. pH

The existence of an ionizable group in the cap structure with pK_a in the physiological range was pointed out above (Sect. B.II.1). Dissociation of the N-1 proton produces the *enolate* form of 7-methylguanosine which is radically different from the *keto* form (Fig. 2), both in net charge and in distribution of electrons. Considering the strict structural requirements for cap activity (Sect. C.IV), it seems unlikely that both forms are recognized equally. We report here experiments to test which form is active. In Figure 8 the pH of the reaction mixture was varied, and translation was measured in the presence and absence of the cap analogue m^7GTP. At low pH, the cap analogue was inactive, but as the pH was raised, it became maximally inhibitory. (The final degree of inhibition is a function of the concentration of cap analogue chosen for the experiment.)

This effect may be due to (a) the *keto* form of the cap being unable to compete for mRNA binding sites or (b) the *keto* form of the cap itself within mRNA being translated by a "cap-independent pathway". To determine whether the latter explanation applied, mRNA was translated at different pH's, but the cap analogue concentrations was increased until maximal inhibition was reached (Fig. 9). Thus, regardless of the ionic form of the cap analogue, the sites to which it bound would become saturated. As can be seen, the overall degree to which mRNA translation could be inhibited increased with higher pH. Thus, the second explanation may account for at least some of the effect.

With such a large number of polypeptides and RNAs involved in protein synthesis, it is impossible to assign the effects on cap recognition of such a general property as pH to a single ionizable

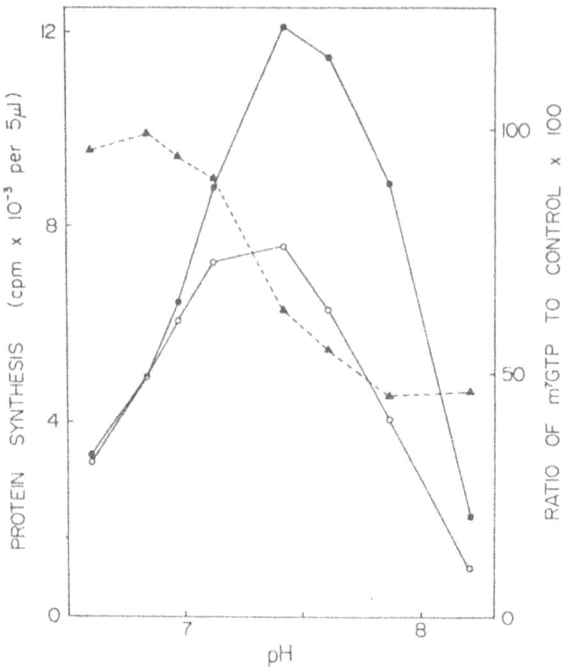

Fig. 8. Translation of globin mRNA in the presence and absence of m^{7}GTP as a function of pH. The cell-free system was that described by Chu and Rhoads (1980). HEPES buffer was used to adjust the pH of the translation system. o-o, translation in the presence of 80 µM m^{7}GTP; ●-●, translation in the absence of m^{7}GTP; ▲-▲, ratio × 100 of translation in the presence of m^{7}GTP to that in its absence. This experiment as well as that in Fig. 9 was performed in collaboration with G. Hellmann, University of Kentucky

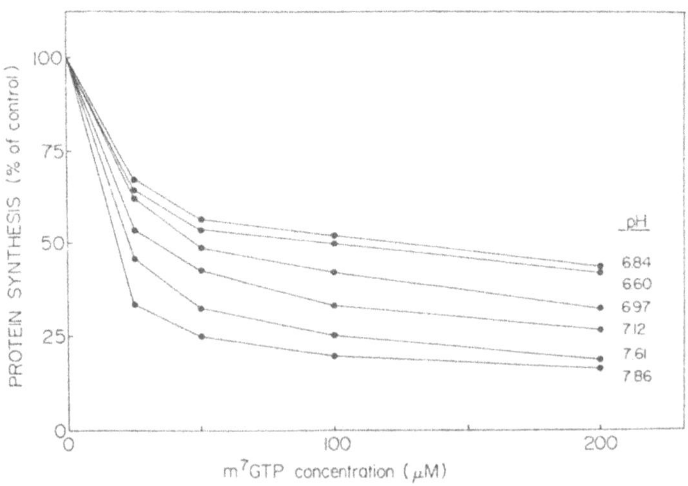

Fig. 9. Inhibition of translation by m^{7}GTP at various pH values. The indicated concentrations of m^{7}GTP were used to inhibit the translation system. HEPES buffer was used to adjust the pH of the reaction mixture to the values shown in the figure

group with certainty. For example, the binding site on the factor which recognizes caps may itself contain an ionizable group with similar pK$_a$. Nonetheless, the coincidence of the pK$_a$ of the N-1 proton of m^7GTP and the pH range over which cap recognition changes is compelling. Further support for the possibility that the *enolate* form is the active one is the fact that $m_2^{1,7}$GDP and Cl^6m^7GDP fail to inhibit protein synthesis, as discussed above (Sect. C.IV.3).

4. mRNA Concentration

A number of investigators, beginning with Shafritz et al. (1976), observed that translational systems made from whole reticulocyte lysate are less sensitive to cap analogue inhibition than those depleted of endogenous mRNA. Various explanations have been proposed. An obvious one is that, since cap analogues inhibit an early step in initiation (Sect. C.II), completion of chains by endogenous mRNA would not be affected while translation of exogenous mRNA, all of which is dependent on initiation, would be completely susceptible. However, since a typical reticulocyte system translates messengers 20 or more times in a standard incubation period (60 min), this would be expected to have only a small effect. Wu et al. (1977) and Asselbergs et al. (1978a) proposed that a cap-recognizing factor is tightly bound to endogenous mRNA and does not detach after initiation; thus endogenous mRNA is less sensitive to inhibition by cap analogues. Willems et al. (1979) suggested that depletion of endogenous mRNA by nuclease treatment decreases the supply of functional initiation factors and leads to "weak initiation complexes" which may be attacked by cap analogues. Chu and Rhoads (1980) proposed a simpler explanation, that of competitive inhibition; the mRNA concentration in untreated lysate is so high that correspondingly high concentrations of cap analogue must be used to adequately compete with it for binding to a cap-recognizing factor. To show this, they restored the globin mRNA concentration in a micrococcal nuclease-depleted lysate (Pelham and Jackson 1976) to the concentration in a whole lysate system. This restored system was again insensitive to cap analogue inhibition. It was also possible to calculate the association constant for mRNA relative to m^7GTP for binding to the putative cap recognizing factor; that of mRNA was approx. 1000-fold higher.

5. Type of mRNA

Numerous reports have appeared documenting the fact that different capped mRNA's are inhibited by cap analogues to different degrees (Suzuki 1976; Lodish and Rose 1977; Asselbergs et al. 1978b; Weber et al. 1978; Sasavage et al. 1979; Willems et al. 1979; Chu and Rhoads 1980). To date, no all-encompassing hypothesis has been advanced to explain these varying sensitivities, with the possible exception of its being due to mRNA secondary structure (Sect. E.III.3). The phenomenon is possibly related to the overall question of mRNA efficiency (reviewed in Lodish 1976; Koch et al. 1980; Walden and Thach 1982). Translational efficiency varies over a range of two orders of magnitude from the "weakest" to the "strongest" mRNA. An intriguing observation, but one which unfortunately does not give direct insight into the problem, is that synthetic capped polynucleotides vary in their affinity for ribosomes from nearly complete binding to almost none, depending on the base composition of the polymer (Sect. C.I.1). Understanding the role of mRNA structure will require further understanding of the mechanism of initiation.

6. Other Factors

In addition to the variables listed above, a number of less well-defined factors have been cited as influencing cap dependence of mRNA translation. These include the source of the cell-free system, the wheat germ and reticulocyte systems being the most frequently compared (Muthukrishnan et al. 1976; Rose and Lodish 1976; Lodish and Rose 1977), and the level of initiation factors (Held et al. 1977; Willems et al. 1979). Some of these differences have been shown to result from the variables cited above (Weber et al. 1978; Chu and Rhoads 1978). Others must await elucidation of the pathway and components responsible for the recognition and positioning of mRNA on the ribosome.

D. Identification of the Cap-binding Protein

The foregoing discussion has served to document the role of the cap structure in translation. It is not an absolute requirement, but it can stimulate the rate of initiation by as much as tenfold, depending on various factors. The step at which this stimulation occurs is binding of mRNA to the 43S initiation complex on the way to its conversion to an 80S complex. Such a role for the cap structure implies that there is an element within the initiation machinery which has affinity for caps but not for unmethylated 5'-termini. Also, the competitive and saturable relationship between cap analogues and mRNA is characteristic of the existence of a binding site. Several independent approaches have been taken to discover this factor, the cap-binding protein. This section will describe these approaches and, simultaneously, present the various ways that the protein may be assayed. Next will be presented methods for purification of the protein.

I. Assays for Cap-binding Protein

1. Cross-linking to Periodate-oxidized mRNA

As discussed above (Sect. B.II.4), the 2'- and 3'-hydroxyls of ribose are converted to aldehydes by $NaIO_4$. Sonenberg and Shatkin (1977) treated reovirus mRNA, labeled with [^{3}H]methyl groups, in this way and incubated it in wheat germ or reticulocyte cell-free systems. Sparsomycin was included as an elongation inhibitor so that the mRNA would be arrested in 48S or 80S initiation complexes. Following incubation, the reaction mixture was treated with $NaBH_3CN$ to reduce any Shiff bases formed to stable cross-links. The remainder of the mRNA was digested with ribonuclease and the radioactively labeled proteins were displayed by gel electrophoresis. When the wheat germ system was used, a number of proteins were crosslinked, ranging from 26 to 135 kd (kilodalton). To determine which were interacting specifically, cap analogues were used to compete for binding. Polypeptides of 26, 93, and 135 kd were eliminated, with concommitant increases at 33 and 100 kd. Purified 48S and 80S initiation complexes contained only the 26-, 93-, and 135-kd polypeptides and not the other nonspecifically bound polypeptides. The same experiment with the reticulocyte system yielded specific binding to proteins of 160 and 35 kd with minor amounts of 135- and 26-kd species.

A later study by Sonenberg et al. (1978) analyzed purified initiation factors from reticulocyte and ascites cells by the same procedure. In total initiation factors, polypeptides of 24 kd and 50 kd were cross-linked, but only that of the 24 kd was sensitive to competition with cap analogue. The same 24-kd polypeptide was found in the initiation factors eIF-3 and eIF-4B (see Sect. E.I.1) but not others. It appeared to be present in substochiometric amounts, however, and thus may not be an integral part of either factor as much as a co-purifying species. Several of the polypeptides in eIF-3 which were cross-linked but not inhibited by cap analogue (160 kd, 135 kd, 60 kd) were the same as those crosslinked in 48S initiation complexes (Sonenberg et al. 1979a). Thus, eIF-3 may be near the site of cap attachment.

2. Reversal of Poliovirus Inhibition

The effects of viruses on initiation of translation with capped mRNA are discussed in detail below (Sect. G). One of the assays for CBP, however, depends on the action of poliovirus on the host cell. Therefore a brief description is included here.

One of the first actions of poliovirus upon infection of HeLa cells is the inhibition of host protein synthesis at the level of initiation (reviewed by Ehrenfeld 1982). Rose et al. (1978) and Helentjaris and Ehrenfeld (1978) observed that cell-free extracts of infected HeLa cells were able to translate poliovirus RNA but not capped mRNA such as that of VSV, while extracts of uninfected cells could translate both. Trachsel et al. (1980) took advantage of this to set up an assay for the component which was altered. Protein fractions from uninfected cells produced a linear increase in the translation of VSV mRNA if the cell-free system was prepared from infected HeLa cells. This assay ultimately led to purification of a 24-kd polypeptide apparently identical to that described in the previous section. As will be discussed below, however, it is not certain that the 24-kd protein is the one responsible for reversal of poliovirus-induced inhibition (Sect. G.I).

3. Reversal of Cap Analogue Inhibition

A third method of detecting cap-binding protein activity is by reversal of inhibition of a cell-free translational system by cap analogues. The rationale of this approach is that cap analogues inhibit protein synthesis by reversibly blocking a site on the protein responsible for recognizing the cap of mRNA. If more

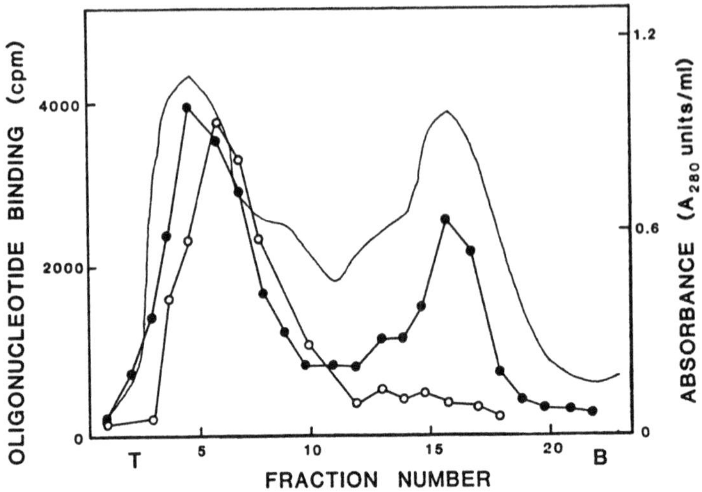

Fig. 10. Detection of cap-binding protein after centrifugation through sucrose gradient in low salt. Rabbit reticulocyte high salt ribosomal extract was purified through Step IV of the CBP purification procedure (see Hellmann et al. 1982). Instead of the cap analogue reversal assay, binding of capped ^{32}P-labeled oligonucleotides was used to detect activity. Binding was either in the absence (•) or presence (o) of 200 μM m^{7}GTP as competitor of nonspecific interactions

of this protein is added, the inhibition is reversed. Hellmann et al. (1982) demonstrated that the high salt wash of reticulocyte ribosomes contained such an activity. By following the activity through fractionation procedures which separate the known initiation factors, a 24-kd polypeptide was isolated which had electrophoretic mobility identical to the one which reacted in the cross-linking assay (Sect. D.I.1) and poliovirus-inhibition-reversal assay (Sect. D.I.2).

4. Binding of Capped Oligonucleotides

Another method of assaying CBP is by directly forming reversible complexes with capped oligonucleotides. Before the CBP had been identified, Both et al. (1975b) demonstrated that 5'-terminal fragments of reovirus mRNA of 32 - 36 nucleotides in length would bind to 80S ribosomes, but those of 7 nucleotides would not. Filipowicz et al. (1976) obtained a salt wash fraction of *Artemia salina* ribosomes which would bind [^{3}H]m^{7}GpppGpC, but purified reticulocyte initiation factors failed to bind under the same conditions. It is not known at this time what protein in the *Artemia* extract was responsible for binding or whether it was the same as the CBP discussed above. Hellmann et al. (1982) purified reticulocyte ribosomal salt wash using the cap analogue reversal assay and showed that fractions which were enriched in the CBP could bind ligands of the form m^{7}Gppp(Np)$_{6-8}$G[^{32}P]pCp. Complex formation was demonstrated by Sephadex G-50 chromatography or retention on nitrocellulose filters. Specificity was shown by the ability of cap analogues to inhibit binding and by the loss of binding upon removal of the cap with tobacco acid pyrophosphatase. Kroath and Shatkin (1982) used a similar approach to detect both the cellular and influenza CBP. Protein was applied to a nitrocellulose filter followed by m^7GpppGmpC[^{32}]pCp. After washing, the bound oligonucleotide was detected by autoradiography.

It is especially important in this type of assay to control for nonspecific binding since many proteins have affinity for RNA or even polyanions in general. This is shown in Figure 10, where crude initiation factors from reticulocyte were sedimented on a sucrose gradient. Binding of oligonucleotides was measured in the presence *(open circles)* or absence *(closed circles)* of cap analogue. CBP is known to associate with eIF-3, the fast-sedimenting peak in the figure, under these conditions (see next section). Binding was observed with both slow and fast sedimenting peaks, but only binding to the eIF-3 peak could be inhibited with cap analogues.

II. Purification of Cap-binding Protein

Several approaches have been taken for purification of CBP. One is to follow one or another of its activities through conventional methods of protein purification. Another is by affinity chromatography using immobilized derivatives of m^7G. Both of these will be discussed in this section. A third approach, binding to antibodies coupled to solid supports, will be presented in Section F.

1. Conventional Methods

Two of the assays listed above, reversal of poliovirus inhibition (Trachsel et al. 1980) and reversal of cap analogue inhibition (Hellmann et al. 1982), have been used to follow activity during fractionation of reticulocyte ribosomal salt wash. In both cases results were similar. Protein was precipitated with 40% saturation of ammonium sulfate. In one case, DEAE-cellulose was used batchwise, and activity eluted between 100 and 210 mM KCl. This was followed, in both schemes, with sucrose gradient sedimentation in 100 mM KCl, in which activity partitioned with eIF-3 (15 - 18S). The centrifugation was repeated, but this time in 500 mM KCl, and under these conditions the activity separated from eIF-3 and sedimented in a broad peak of 8 - 12S. The material was applied to DEAE cellulose, and activity was eluted at approximately 150 mM KCl. At this stage the preparation consisted of polypeptides of 24, 50, 90, and 110 kd. The final step was to apply it to phosphocellulose and elute with 400 mM KCl. Both preparations exhibited the appropriate activity at this stage and contained only the CBP.

A troublesome aspect noted in both studies is that highly purified CBP lost activity rapidly, usually overnight. Less purified fractions were stable to both manipulation and storage. Trachsel

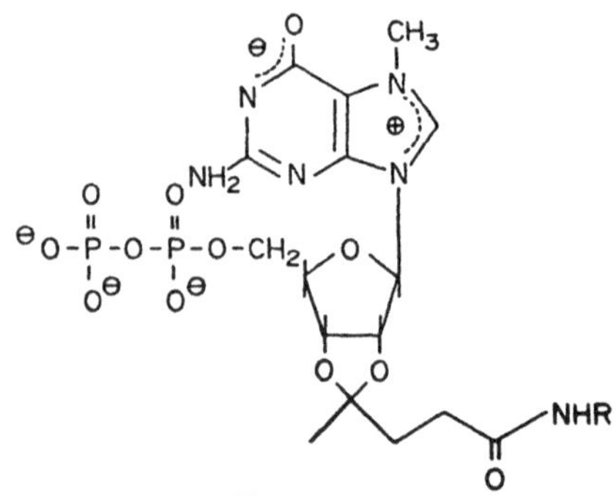

5

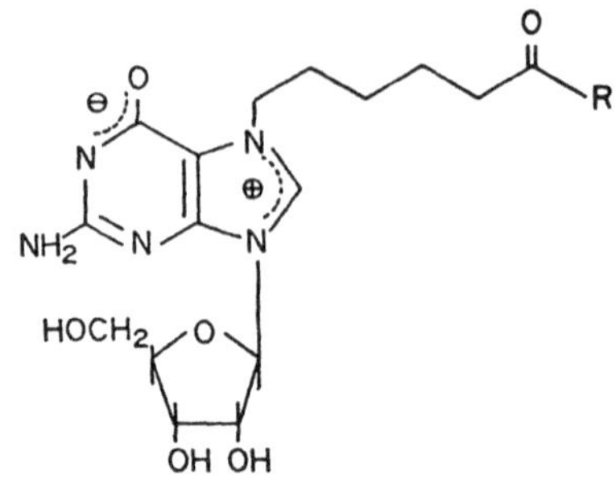

6

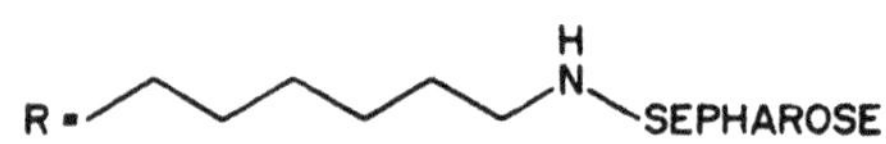

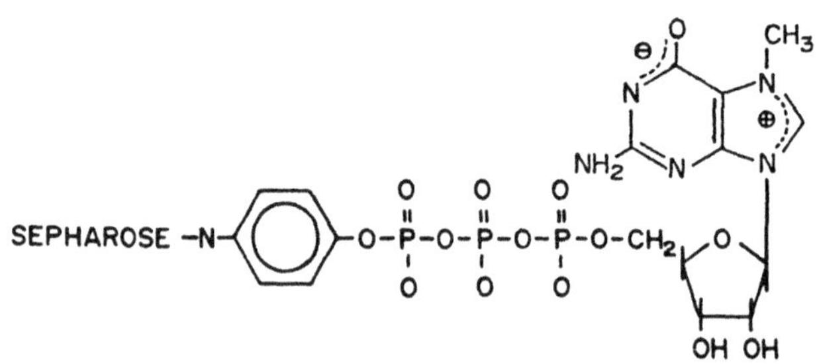

7

Fig. 11. Affinity media used for the purification of CBP. 5 2',3'-O-[1-(2-carboxyethyl)ethylidene]-7-methylguanosine 5'-diphosphate linked to Sepharose (Sonenberg et al. 1979b). 6 7-(5-carboxypentyl)guanosine 5'-diphosphate linked to Sepharose (Rupprecht et al. 1981). 8 the p-nitrophenyl-γ-ester of m^7GTP linked to Sepharose. (Webb et al. 1984)

et al. (1980) reported that cruder fractions could reactivate CBP which had lost the ability to restore poliovirus infected lysates, suggesting that additional components were required to maintain activity.

2. Affinity Methods

The elucidation of the structural features of the cap which are necessary for its function (Sect. C.IV) has made it possible to design immobilized ligands which have affinity for CBP. Three separate approaches have been taken.

Sonenberg et al. (1979b) prepared the levulinic acid $O^{2',3'}$-acetal of m^7GDP (Fig. 11, compound 5) and coupled it to AH-Sepharose 4B.

The material applied to the column was reticulocyte ribosomal salt wash which had been fractionated with 0 - 40% saturation of ammonium sulfate and centrifuged on glycerol gradients at 0.1 M KCl. Only the upper portion (non-eIF-3) was used for affinity chromatography. Most proteins did not bind, and those which did could be eluted with 1 M KCl. The eluted proteins included a large number of high mol. wt. species, but the CBP was also among them in minor quantity. Nearly the same spectrum of polypeptides was retained by the control resin GDP-Sepharose, the exception being the 24-kd polypeptide. However, the cross-linking assay (Sect. D.I.1) revealed that m^7GDP-Sepharose specifically retained CBP. The specificity of the procedure was improved by eluting with 70 μM m^7GDP instead of 1 M KCl. This yielded only CBP.

Rupprecht et al. (1981) synthesized a different affinity ligand, based on the prior observation (Sect. C.IV.3) that the alkyl substituent at position 7 of the guanine ring was not restricted to a methyl group. The compound, 7-(5-carboxypentyl)guanosine 5'-diphosphate was coupled to AH-Sepharose 4B via the carboxyl group (Fig. 11, compound 6). The same initiation factor preparation was applied as described above, and after washing in 0.1 M KCl, protein was eluted first with 70 μM m^7GDP and then with 1 M KCl. Only CBP was eluted with m^7GDP. This was shown to be active by the cross-linking assay and by differential stimulation of capped (Sindbis virus, reovirus, globin) mRNA but not uncapped (EMC virus, STNV) RNA. A control ligand lacking the 5'-diphosphate was also synthesized and shown to have no affinity for CBP, as expected (Sect. C.IV.1).

A third approach has been taken by our own laboratory in collaboration with Dr. John Kozarich, Yale University. The γ-phosphate of m^7GTP was coupled to p-nitrophenol and this derivative, after reduction to the aniline compound, was linked to cyanogen bromide-activated Sepharose 4B (Fig. 11, compound 7). Unfractionated reticulocyte ribosomal salt wash was applied to this material and protein eluted with 70 μM m^7GTP. Again, only CBP was eluted (Fig. 12). Activity was demonstrated by the oligonucleotide binding assay (Sect. D.I.4). Stripping the column with sodium dodecylsulfate revealed that all of the CBP had been eluted by m^7GTP.

In comparing these three affinity media, one may note that synthesis of compound 7 is the easiest and most straightforward. The ligand perhaps also has the highest affinity for CBP, based on the following considerations. From a theoretical standpoint, compound 7 is the most similar to the natural cap of mRNA. If the conformation of the ribose ring is indeed important for cap function (Sect. B.II.3), then derivatization of the 2',3'-diol in 5 could affect this. Similarly, even though the 7-methyl group can be replaced with ethyl or benzyl, it is nonetheless very close to regions of the guanine ring which must be in direct contact with the protein. A practical indication that binding may be tighter is that unfractionated initiation factors may successfully be used with 7 (Fig. 12). Sonenberg et al. (1979b), on the other hand, state that if reticulocyte ribosomal salt wash from which eIF-3 has not been removed, i.e., the 0 - 40% saturation ammonium sulfate cut, was applied to 5, no polypeptides were

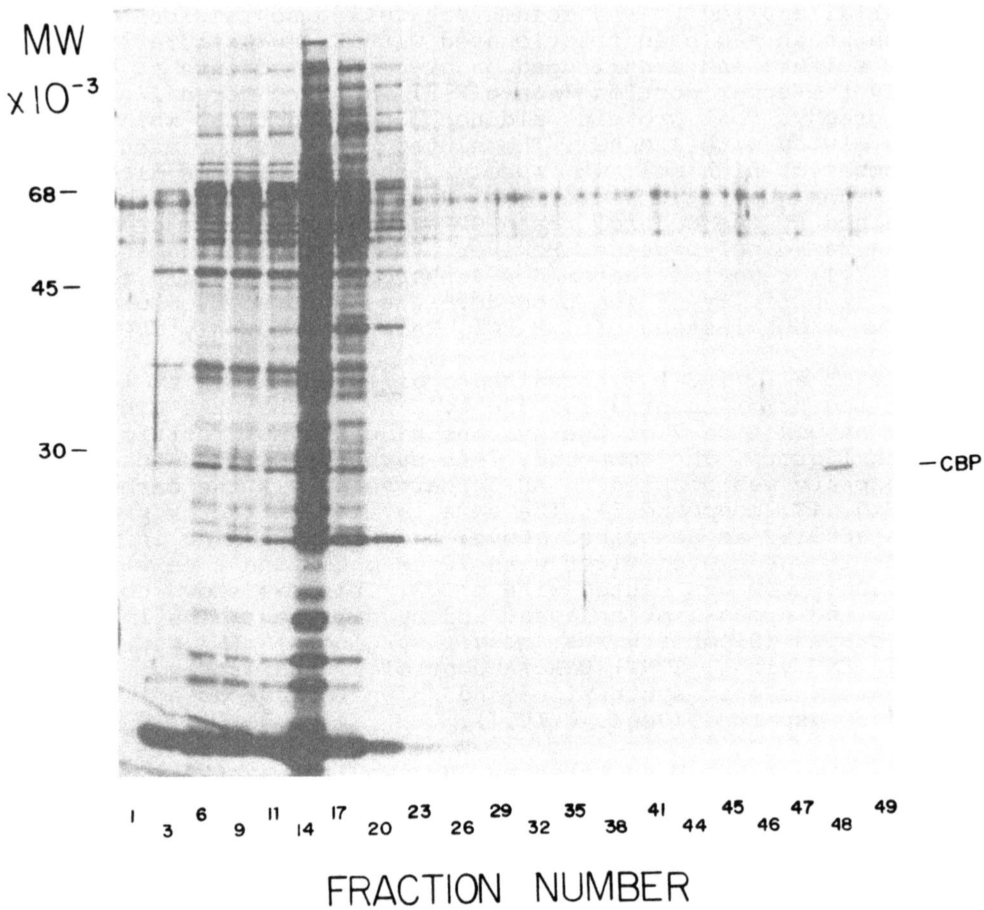

Fig. 12. Purification of CBP using the p-nitrophenyl-γ-ester of m^7GTP linked to Sepharose. A 0-70% saturation ammonium sulfate cut of rabbit reticulocyte ribosomal salt wash (17 A_{280} units) was applied to a 1-ml affinity column in B-100 buffer (Hellmann et al. 1982). Fractions of 0.5 ml were collected and aliquots analyzed by denaturing gel electrophoresis with visualization by silver staining. Numbers designate column fractions. Elution with 70 μM m^7GTP was begun at fraction 45. The prominent pair of bands at approx. 55 and 65 kd are artifacts of the staining procedure and not proteins. This experiment was performed in collaboration with N. Webb, University of Kentucky, and J. Kozarich, Yale University

eluted with 70 μM m^7GDP. There may be some lack of specificity with binding to 8 as well. Rupprecht et al. (1981) showed that, while m^7GDP was able to elute some of the CBP (as assayed by cross-linking), the bulk of the activity remained bound and was eluted only with 1 M KCl.

E. Involvement of Cap-binding Protein in Initiation of Protein Synthesis

This section deals with the mechanism by which CBP recognizes messenger RNA and its relationship to other components involved in the initiation of protein synthesis. Unfortunately, the molecular details are not known at this time, and this is currently a subject of considerable research effort. Nonetheless, the various pieces of the puzzle will be reviewed.

I. Overview of the Initiation of Protein Synthesis

1. Roles of the Initiation Factors

Several articles have appeared which describe each step occurring during initiation and the factors catalyzing them (Trachsel et al. 1977; Benne and Hershey 1978; Jagus et al. 1981). The following is a brief "consensus" description. The reader is referred to the individual articles for a complete review of the evidence supporting it.

Whereas most workers refer to initiation complexes involving the small ribosomal subunit as 40S initiation complexes, Jagus et al. (1981) conveniently distinguish the various steps in initiation complex formation by sedimentation properties: 43S ribosomal subunits contain 40S ribosomal subunits and eIF-3; 43S initiation complexes contain 40S ribosomal subunits, eIF-3, met-tRNA$_f$ and GTP; 48S initiation complexes contain mRNA as well. This nomenclature will be used here.

The 40S and 60S ribosomal subunits have a high affinity for each other. Formation of an inactive 74S couple is prevented by interaction of eIF-3 with the 40S subunit to make a 43S subunit. Another factor which is involved at this stage is eIF-4C. The second step is formation of a ternary complex of eIF-2, GTP and met-tRNA$_f$. This complex joins the 43S subunit to make a 43S initiation complex. The next step is addition of mRNA to form a 48S initiation complex. The most important initiation factors for this step are eIF-3, eIF-4A and eIF-4B. Other factors which stimulate mRNA binding, although not all groups are in agreement, are eIF-1 and eIF-4C. The 40S subunit becomes positioned at the intiation codon (see next section) and the 60S subunit joins, accompanied by release of eIF-2 and eIF-3. This last step requires eIF-5.

2. Selection of the Initiation Codon

Baralle and Brownlee (1978) surveyed the known 5'-terminal sequences of eukaryotic mRNA's and observed that the only common sequence was the initiator AUG. Kozak and Shatkin (1978) proposed a scanning model for initiation codon selection in which the 43S initiation complex binds first to the cap and then migrates down the mRNA until the first AUG is found. The considerable evidence supporting this model is summarized by Kozak (1978). This, then, represent a major, if not the fundamental, difference between eukaryotic and prokaryotic protein synthesis.

3. The Involvement of ATP in Initiation

The binding of mRNA has been known since 1970 (Marcus 1970a,b) to require the hydrolysis of ATP, but the precise step has not been found. It has been narrowed down to the reaction between the 43S initiation complex and mRNA (Trachsel et al. 1977; Benne and Hershey 1978), which is also the site of inhibition by cap analogues (Section C.II). Sonenberg and Shatkin (1977) used the nonhydrolyzable analogue AMP-P(NH)P and showed that it decreased cross-linking of caps to the 26- and 135-kd polypeptides in wheat germ initiation complexes, suggesting that the requirement may be at the cap recognition step. Sonenberg (1981) subsequently carried out experiments on the effect of ATP and Mg^{++} on cross-linking of caps in a partially purified preparation of reticulocyte initiation factors (not containing eIF-3). He confirmed that ATP decreased cross-linking to CBP but extended this to show that new bands at 28 and 50 kd appeared. In a later publication, Sonenberg et al. (1981a) clearly demonstrated that an 80-kd polypeptide is also cross-linked in the presence of ATP and Mg^{++}, together with the 28-kd and 50-kd polypeptides. Based on subsequent studies, the 80 and 50-kd polypeptides are likely to be eIF-4B and eIF-4A, respectively.

In spite of the alteration in the pattern of polypeptides cross-linked, ATP is not essential for direct complex formation between capped oligonucleotides and purified CBP. Hellmann et al. (1982) showed that specific binding could occur in the absence of ATP. It was stimulated approx. two-fold, however, by $MgCl_2$ and ATP, GTP, or CTP. These apparently contradictory findings, that ATP is involved at the cap recognition step but does not specifically affect complex formation with purified CBP, has been clarified considerably by a recent study by Grifo et al. (1982), who showed that cross-linking of mRNA caps to CBP was stimulated by eIF-4A. This polypeptide (44 kd) was also itself cross-linked to mRNA, as was eIF-4B (80 kd) when present in incubation mixtures. It is likely that these are the same 50-kd and 80-kd polypeptides identified earlier by Sonenberg (1981). Cross-linking to both of these latter polypeptides was stimulated by ATP and inhibited by m^7GDP. However, cross-linking to CBP was not stimulated by ATP. Also, if either eIF-4A or eIF-4B were tested alone, ATP had no effect on cross-linking. The interpretation of these results is not entirely straightforward at this time, but it appears that eIF-4A is the ATP-requiring protein. The fact that cross-linking to it requires CBP suggests that the two factors act together. EIF-4B may also be involved, but that is not yet certain since eIF-4B preparations which are free of CBP are not yet available. The results also open the possibility that cap binding by CBP occurs prior to binding to the ribosome in an ATP-independent step, but then is followed by transfer of the mRNA together with eIF-4A and eIF-4B to the 43S initiation complex coincident with ATP hydrolysis.

Another process during initiation of protein synthesis which appears to require ATP is 43S migration. Kozak (1980a) has investigated this by loading reovirus mRNA with 43S ribosomal subunits in the presence of edeine, which prevents recognition of AUG codons and allows the 43S subunit to reach the 3'-end of the mRNA. These were shown to run off only if ATP was present. Also, depletion of ATP by adding hexokinase and glucose prevented 43S subunits from reaching the 3'-end of the mRNA. This requirement for ATP is in addition to the mRNA binding step: 43S ribosomes which were bound to mRNA required further hydrolysis of ATP to reach the first AUG codon. Nonetheless, the requirement during mRNA binding could be related mechanistically to the migration process. For instance, ATP-independent cap binding by CBP may occur on the 43S subunit and be immediately followed by ATP-dependent 40S migration catalyzed by eIF-4A to generate a stable complex.

II. Association of Cap-binding Protein with the Known Initiation Factors

It is reasonable to suppose that many of the initiation factors perform their functions in close physical proximity and cooperate mechanistically with successive steps in the initiation process. Even though these relationships have not yet been worked out, a number of observations give hints about the overall organization. In this section, the physical relationship and co-purification of CBP with other factors will be discussed.

1. eIF-3

This is by far the largest and most complex of the initiation factors, and this has led to some disagreement in the literature as to the mol. wt. and number of polypeptides which are integral components. For the rabbit reticulocyte factor, the number of polypeptides ranges from seven to eleven and the mol. wt. from 400 to 700 kd (Trachsel et al. 1977; Benne and Hershey 1978; Jagus et al. 1981). The wheat germ (Checkley et al. 1981) and HeLa cells (Etchison et al. 1982) factors are similar to this. As mentioned above (Sect. E.I.1), eIF-3 is present as part of the 43S ribosomal subunit prior to mRNA binding, is required for mRNA binding, and dissociates after the 40S subunit has located the first AUG codon, coincident with 60S subunit joining. Thus, it is a natural candidate to be involved with CBP.

The smallest polypeptide detected among the constituents of eIF-3, but only by some investigators, is 24 kd. This is likely to be CBP, based on functional studies. As mentioned above (Sect. D.II.1), CBP co-purifies with eIF-3, whether activity is being followed by capped oligonucleotide binding, poliovirus inhibition reversal, or cap analogue inhibition reversal, up through the sucrose gradient step in low salt. At this stage, all of the known initiation factors have been separated from eIF-3. CBP activity is then separated from eIF-3 on a sucrose gradient in high salt. The cross-linking assay has also been used to show that CBP is found in eIF-3 preparations (Sonenberg et al. 1978; Bergmann et al. 1979). Furthermore, Sonenberg et al. (1979a) compared the polypeptides other than CBP which were positive in the cross-linking assay with the polypeptide subunits of eIF-3. The results indicated that the 160-kd, 130-dk, 60-kd, and 35-kd polypeptides could all cross-link mRNA caps. Hansen et al. (1982b) showed that most of the cross-linking activity to the 24-kd polypeptide was associated with eIF-3. Activity was also found, however, in the 200,000 × g supernatant fraction in a 5 - 7S form (Hansen et al. 1982a).

Another major line of evidence that CBP is associated with eIF-3 comes from studies with poliovirus-infected HeLa cells. This will be considered in detail in Section G. For the present, suffice it to say that only eIF-3 is effective in restoring the ability to translate capped mRNA to lysates from infected cells (Helentjaris et al. 1979; Trachsel et al. 1980). CBP is associated with eIF-3 from uninfected cells but is absent from that of infected cells.

Taken together, all of these results indicate that there is a physical association between eIF-3 and CBP. What is not yet certain, however, is whether this association is a result of the high salt treatment which is used in all purification schemes for initiation factors (or some other treatment) or is a true physiological association. The cross-linking suggests a specific association, since only four of the eleven or so polypeptides of eIF-3 are cross-linked to mRNA. These polypeptides presumably do not represent alternate sites for cap binding but rather are polypeptides located near the site of attachment on eIF-3 for CBP.

2. eIF-4B

This initiation factor is apparently a single polypeptide of 80 kd although it has been difficult to obtain in homogeneous form. It is required for mRNA binding and comes into play only after the 43S initiation complex has been assembled. It was originally claimed to be the factor responsible for recognizing the mRNA cap since purified preparations of eIF-4B (a) bound to capped but not uncapped mRNA's in a specific (cap analogue inhibitable) fashion, as assayed by retention on nictrocellulose filters (Shafritz et al. 1976; Baglioni et al. 1978); (b) reversed inhibition of cell-free translation by cap analogues (Padilla et al. 1978); and (c) restored the ability to translate capped mRNA to lysates of infected HeLa cells (Rose et al. 1978). It is now generally agreed that eIF-4B does not specifically recognize mRNA caps. Several authors have proposed explanations for the data to the contrary (Sonenberg and Shatkin 1978; Jagus et al. 1981; Hellmann et al. 1982). In addition, preparations of eIF-4B appear to contain trace amounts of CBP, as assayed by cross-linking to mRNA (Sonenberg et al. 1978; Bergmann et al. 1979). Nonetheless, fractionation of initiation factors using as an assay either cap analogue inhibition reversal (Hellmann et al. 1982) or poliovirus inhibition reversal (Helentjaris et al. 1979; Trachsel et al. 1980) did not yield eIF-4B.

Considering the stimulatory role of eIF-4B in cross-linking of mRNA to CBP, eIF-4A, and eIF-4B itself (Sect. E.I.3), it seems reasonalbe to suppose that CBP and eIF-4B have some affinity for each other. They may also cooperate in catalyzing a given step or successive step in initiation, so that the presence of one factor would amplify detection of the other in some types of assay systems.

3. eIF-4A

This factor consists of a single polypeptide, variously assigned molecular weights from 44 to 50 kd, which is required for the binding of mRNA to the 43S initiation complex (see references in Grifo et al. 1982). Its ability to be cross-linked to mRNA and to catalyze ATP-dependent cross-linking to eIF-4B and CBP was discussed above (Sect. E.I.3). It is likely to be the 50-kd species which Sonenberg (1981) identified as being cross-linked to mRNA caps only in the presence of ATP and Mg^{++}. An activity which has not previously been discussed is the ability of a factor to relieve competition between different mRNA's for translation. Ray et al. (1983) studied the competition between reovirus and globin mRNA and showed that, of all the known initiation factors, only eIF-4A and a complex containing eIF-4A were able to relieve the competition. The composition and role of complexes containing eIF-4A and CBP will be considered below (Sect. E.II.5).

4. eIF-2

This factor consists of three polypeptides and has clearly been shown to be responsible for forming a ternary complex with GTP and met-$tRNA_f$, followed by binding to the 43S ribosomal subunit (reviewed in Jagus et al. 1981). Cross-linking of the 48- and/or 52-kd polypeptides of eIF-2 to mRNA occurs, but it is not inhibited by m^7GDP (Sonenberg et al. 1978). Kaempfer et al. (1978) presented data that eIF-2 reversed inhibition of translation by cap analogues (although Hellmann et al. 1982, did not find this). Also, Kaempfer et al. (1978) showed that cap analogues inhibited binding of either met-$tRNA_f$, capped mRNA, or uncapped mRNA to eIF-2. Sonenberg and Shatkin (1979) extended these results to include 18S ribosomal RNA and raised the question of whether inhibition by cap analogues was a reliable test of cap-specific interactions between mRNA and proteins. To date, CBP has not been found in eIF-2 preparations. Thus, at present, the evidence is not compelling that eIF-2 is involved with cap recognition or CBP.

5. Complexes Containing CBP

Several of the findings revealed up to this point suggest, with the benefit of hindsight, that CBP may exist in complexes with other proteins. These findings include the fact that a variety of proteins are cross-linked to mRNA caps in initiation complexes (Sect. D.I.1), that the CBP co-purifics with eIF-3 and when dissociated, sediments much faster than a protein of 24,000 daltons (Sect. D.II.1), and that a factor such as eIF-4A stimulates not only ATP-dependent cross-linking of mRNA caps to CBP but also to eIF-4A itself and eIF-4B (Sect. E.I.3). Direct evidence for complexes has now been obtained by several investigators. Tahara et al. (1981) sedimented the high salt ribosomal wash of rabbit reticulocytes on a sucrose gradient in 0.5 M KCl and applied various fractions to a m^7GDP-Sepharose affinity column. They observed, in different parts of the gradient, CBP either alone or in combination with various polypeptides (48 kd, 55 kd, 94 kd, 210 kd). This complex was termed CBPII by the authors. The 48 kd polypeptide was shown to have identical migration on denaturing gel electrophoresis as eIF-4A, and like eIF-4A, could be cross-linked to mRNA caps in an ATP-dependent reaction (Tahara et al. 1982). The fractions containing only CBP did not possess poliovirus inhibition restoring activity, but the higher mol. wt. complexes did. Interestingly, the latter complexes were stable to repeating freezing and thawing, unlike the purified CBP (Sect. D.II.1). Thus, the other polypeptides may either stabilize CBP or provide separate activities related to cap recognition which are inactivated by poliovirus.

Another complex has been purified by Grifo et al. (1983) using an entirely different approach. These authors constructed a cell-free system which was deficient in eIF-4B. Crude reticulocyte ribosomal salt wash was fractionated and assayed both for globin synthesis in the reconstructed system and for poliovirus inhibition reversal in HeLa cell extracts. A complex was isolated which possessed both activities, while a preparation consisting primarily of the 80-kd polypeptide of eIF-4B possessed only the globin-synthesis-stimulating activity. The complex contained polypeptides of 24, 46, 73 and 200 kd, similar to CBPII. The authors termed this complex eIF-4F and demonstrated that the reconstituted system has independent requirements for eIF-3, eIF-4B, eIF-4A and the new factor eIF-4F. The 46-kd polypeptide was shown to be eIF-4A by several criteria. The combination of eIF-4A, eIF-4B, and CBP will cause mRNA to be retained on nitrocellulose filters and, as discussed previously (Sect. E.I.3), will cross-link to mRNA caps in an ATP-dependent fashion. The same properties were demonstrated for eIF-4F. Finally, eIF-4F was active in binding mRNA to 43S initiation complexes.

The two complexes described here are very similar to each other in properties and probably represent the same entitiy. The factor eIF-4F exhibits several types of activities, and they are not all reproduced by combining eIF-4A, eIF-4B and CBP. The presence of the 200-kd polypeptide may have some as yet undefined role. eIF-4F, then, may represent the functional form of CBP. Even if this does represent the next level of organization, however, the individual roles of the polypeptides and overall mechanism of action are not yet understood.

III. Effect of Cap-binding Protein on the Translation of Different mRNA's

One way to study the role of CBP is to alter the mRNA being translated. In this section various approaches of this kind are reviewed as well as the insights they give about function.

1. Translation of Uncapped mRNA's

As discussed above (Sect. C.I.1), mRNA's lacking a cap were used prior to the discovery of CBP to demonstrate the fact that translation is stimulated by the presence of a cap. After discovery of CBP, capped and uncapped mRNA's were shown to be differentially stimulated by the protein. Sonenberg et al. (1979b) purified CBP to homogeneity by affinity chromatography (Sect. D.II.2) and tested it with various mRNA's. Translation of the uncapped RNA's of STNV and EMC virus were stimulated less than 20%, while those of globin, reovirus and Sindbis virus, all capped, were stimulated two- to fourfold. The effect of placing both an uncapped and capped mRNA in the same translation mixture was explored using alfalfa mosaic virus 4 and STNV RNA's together (Sonenberg et al. 1980). When the concentration of purified CBP was increased, only the products of the capped RNA were stimulated, while those of the uncapped were slightly decreased. Other combinations of capped and uncapped RNA's gave similar results. Other reticulocyte initiated factors were tested for the same activity. Of these, only eIF-3 and eIF-4B produced differential stimulation, both of which are known to contain small amounts of CBP (Sect. E.II.1-2). If eIF-3 was repurified by sedimentation through sucrose gradients containing 0.5 M KCl to remove the CBP, its ability to discriminate between messengers was lost.

2. Differential Translation of Capped mRNA's

Another question is whether there is translational discrimination between capped mRNA's. A favorite test system has been the mRNA's for α and β globin (reviewed by Lodish 1976). More recently, Kabat and Chappell (1977), showed that eIF-4B would overcome the preferential translation of β over α mRNA. Parets-Soler et al. (1981) used the stimulation of the α/β ratio as an assay for purifying a factor from reticulocyte ribosomal salt wash. They obtained two polypeptides of 28 and 50 kd, which, based on their size and activity, may be CBP and eIF-4A. Further evidence for mRNA discrimination by CBP or its complexes comes from the study of Ray et al. (1983) mentioned previously (Sect. E.II.3). By setting up a competitive situation between reovirus and globin mRNA they were able to demonstrate that only eIF-4A and CBP were active in this respect. While many theories have been advanced to explain these intriguing observations, the biochemical basis for preferential translation of some capped mRNA's over other remains unexplained.

3. Secondary Structure of mRNA

The preceding studies have used natural mRNA's to determine a differential effect of CBP. It has also been possible to modify mRNA's to introduce more or less secondary structure and determine its effect on initiation of protein synthesis. Morgan and Shatkin (1980) used inosine 5'-triphosphate in place of GTP in the reovirus transcription system. Under appropriate conditions, mRNA's could be synthesized which were substituted with I for G internally but not in the cap, and these exhibited considerably less secondary structure (due to the absence of G:C base pairing). Ribosome binding was enhanced and movement of ribosomes on such mRNA was freer. Kozak (1980a) explored the requirement for the cap in both inosine-substituted mRNA and bisulfite-treated mRNA (to convert C to U) and showed that uncapped mRNA of this type could readily be incorporated into 80S ribosomes and polysomes whereas the control (guanosine-containing) mRNA was unable to do so. Thus, the requirement for a cap is obviated to some extent by unfolding of the mRNA. This suggests that the role of the cap may be to facilitate exposure of the 5'-terminus in a highly folded mRNA. Kozak (1980b) also used inosine-substituted mRNA to show that the ATP requirement for formation of 80S initiation complexes could be abolished. Thus, 43S ribosomal subunit migration may require ATP only to melt secondary structure in the mRNA. A direct effect of mRNA secondary structure on cap binding was demonstrated by Lee et al. (1983), who used inosine substitution to decrease secondary structure and bromouridine substitution or elevated KCl concentrations to increase it. They observed that the cross-linking to 28-, 50-, and 80-kd polypeptides, previously shown to require ATP (Sect. E.I.3), occurred in the absence of ATP in mRNA of low secondary structure. mRNA with increased secondary structure required ATP, and optimal cross-linking occurred at lower KCl concentrations than with native mRNA. These results confirm and extend the hypothesis stated above, that ATP-dependent melting of mRNA secondary structure may be involved in the action of cap recognition factors.

F. Antibodies to Cap-binding Protein-related Polypeptides

In view of the large number of polypeptides discussed up to this point which are involved in some way in the entry of mRNA into the cycle of protein synthesis and considering the overall complexity of the system, it would be extremely valuable to have specific antibodies directed toward the various components. Several attempts have been made to develop such antibodies, and their characterization as well as functional activities will be discussed in this section.

I. Anti-eIF-3

As mentioned above (Sect. E.II.1), CBP co-purifies with eIF-3. Sonenberg and Trachsel (1982) injected eIF-3 into mice and fused spleen cells with myeloma cell lines to form hybridomas. The hybridomas were screened with CBP purified by affinity chromatography on m^{7}GDP-Sepharose (Sect. D.II.2). One of the positive clones produced an IgM class of antibody and was termed E.2.4.D_4. When the eIF-3 preparation to which the antibody was made was tested by immunoblotting, polypeptides of 210, 160, 65, 50, 28, and 18 kd reacted, but not CBP. Several of these polypeptides appeared to be related to each other, based on the presence of common tryptic peptides. In particular, the 28- and 24-kd polypeptides were similar, as were the 24 and 50 kd. When reticulocyte mono- and polyribosomes were reacted with the antibody, only the 50-kd species was found. The only reactive species in postribosomal supernatant, on the other hand, was the 210-kd species. An immunoaffinity column was used to purify proteins from the postribosomal supernatant, and this yielded predominantly 210- and 65-kd polypeptides, with smaller amounts of several others.

Functional analysis of the IgM antibody suggested that it recognized proteins involved in the cap recognition process. It inhibited ATP-dependent cross-linking of mRNA caps to the 24-, 28-, 50-, and 80-kd complex (see Sect. E.I.3). Also, the antibody inhibited cell-free translation of (capped) reovirus mRNA but not (uncapped) EMC virus RNA.

These results are intriguing but somewhat difficult to interpret since the monoclonal antibody does not react with only one polypeptide. This may be because the polypeptides are related to each other with the larger ones, perhaps, giving rise proteolytically to the smaller. Alternatively, there may be trivial epitopes shared by a number of unrelated proteins. The presence of similar tryptic peptides in some of the proteins adds weight to the former interpretation. Nonetheless, the fact that reaction with CBP was not detected necessitates caution in interpreting the results.

II. Anti-Heme Controlled Repressor

A second monoclonal antibody was made by injecting a partially purified preparation of the heme-controlled repressor of rabbit reticulocyte into mice and again screening with CBP (Sonenberg and Trachsel 1982; Sonenberg et al. 1981a). A hybridoma producing IgG class antibodies, termed B_3/A_1, was selected. An immunoaffinity column made from this antibody was used with a 0 - 40% saturation ammonium sulfate cut of reticulocyte ribosomal salt wash. The major polypeptide bound was one of 50 kd, and this polypeptide also reacted with the IgM antibody described above.

Functionally this antibody behaved similarly to the IgM antibody, i.e., it inhibited (capped) reovirus mRNA translation but not that of (uncapped) STNV RNA. The binding of reovirus mRNA to 80S ribosomes was also inhibited by the antibody. Interestingly, inosine-substituted mRNA was less sensitive to inhibition, and bromouridine-substituted mRNA more sensitive, than control mRNA. This is in keeping with the hypothesis stated above that cap recognition is more important for translation of mRNA's with more secondary structure (Sect. E.III.3). An unexpected finding with this antibody was that the translation of various capped mRNA's was inhibited to varying degrees. For example, under conditions where translation of globin mRNA was inhibited 95%, that of alfalfa mosaic virus 4, which is known to be capped, was not affected. The degree of inhibition by the antibody correlated with the sensitivity of translation to m^7GDP, suggesting that some mRNA's are less dependent on the cap. This may be related to factors such as secondary structure of the mRNA.

As with the IgM antibody, these studies provide support for existing hypotheses about the mechanism of mRNA entry into the initiation process. However, they are clouded by the uncertainty of what polypeptides they recognize.

III. Anti-24-kd Cap-binding Protein

A third monoclonal antibody was made by injection of CBP, purified from rabbit erythrocytes by affinity chromatography on m^7GDP-Sepharose (Shatkin et al. 1981). The antibody, an IgG class, was shown to inhibit translation of capped Sindbis mRNA but not that of (uncapped) EMC virus. Also, it prevented the stimulation which could be produced by adding affinity purified cap-binding protein. The nature of the polypeptides with which the antibody reacted, however, was not reported.

G. Effect of Virus Infection on Cap-binding Protein

Viruses are entirely dependent upon the host cell for translation of their RNA, whether the RNA be encapsidated or produced by transcription within the host cell. It is not surprising, then, that a number of strategies have evolved which allow viruses more ready access to the host protein synthetic machinery. For at least one virus, polio, and perhaps others, the target for facilitating takeover of host translation is the cap recognition step. This has spurred a great deal of interest in the identity and mechanism of the proteins which are involved in this step. In this section the effects of viral infection on initiation of protein synthesis will be discussed.

I. Poliovirus

Ehrenfeld (1982) has recently reviewed poliovirus-induced inhibition of host-cell protein synthesis. Briefly, infection of HeLa cells gradually inhibits host protein synthesis for up to approximately two hours, with concomitant loss of capped mRNA from polysomes. The cellular mRNA, however, remains functionally intact. At this time, polysomes begin to increase again but contain only uncapped poliovirus mRNA. Cell-free systems made from such infected cells are unable to translate capped mRNA's but readily translate uncapped mRNA's. Initiation factors from uninfected cells can restore the ability to translate capped mRNA's. The search for the defective component was described in Section D.I.2 and eventually led to CBP, although it was originally thought to be first eIF-3 and then eIF-4B.

The implication of this finding is that poliovirus, which is uncapped, somehow inactivates CBP and thereby facilitates translation of its own RNA by inhibiting that of the host. Unfortunately, things are not so simple. Hansen and Ehrenfeld (1981) found that CBP from infected cells was fully active in the cross-linking assay. Similarly, Tahara et al. (1981) showed that CBP alone, purified by affinity chromatography, was inactive in restoring activity to infected cell lysates, but complexes containing CBP and larger polypeptides (see Sect. E.II.5) were active. Also, eIF-4F, but not CBP, had poliovirus-inhibition-reversal activity (Grifo et al. 1983). Thus, CBP *per se* is not the affected component. It should be pointed out that there is some disagreement in the literature concerning this point. Lee and Sonenberg (1982) have presented results which indicate that cross-linking to CBP and associated 28-, 50-, and 80-kd polypeptides is reduced after poliovirus infection, in contrast to the results of Hansen and Ehrenfeld (1981). This point is not yet resolved.

New insight into this problem came from a study by Hansen et al. (1982), who studied the association between CBP and eIF-3 in normal and poliovirus-infected HeLa cells. They found that, whereas these two factors form a complex normally (Sect. D.II.1), in infected cells, CBP was completely absent from eIF-3, as detected by the cross-linking assay. In a striking finding, perhaps responsible for the above result, Etchison et al. (1982) reported that poliovirus infection is accompanied by proteolysis of a 220-kd polypeptide in eIF-3. This polypeptide is reminiscent in size of the 200-kd components of CBPII and eIF-4F (Sect. E.II.5) and the 210-kd component detected with Trachsel's monoclonal antibodies (Sect. F.I). The polypeptide was associated with eIF-3 but could be removed in high salt. Antibody against this component reacted with a polypeptide of the same size in complexes of cap-binding protein purified by affinity chromatography on m^7GTP-agarose.

All of these results, while very informative, still do not indicate how poliovirus RNA is able to be translated in infected cell lysates but capped RNA is not. It is not simply the absence of a cap, because VSV mRNA from which the cap has been chemically removed is not translated either. An insight into this has come from the finding of Sonenberg et al. (1982) that capped mRNA's with reduced secondary structure (Sect. E.III.3) are able to bind ribosomes in infected cell extracts. Thus, there may be features of poliovirus RNA other than the absence of a cap which enable it to be translated by the altered initiation machinery. One of these may be an unusually low affinity or requirement for eIF-4A (Daniels-McQueen et al. 1983).

While the ultimate mechanism of inhibition of protein synthesis by poliovirus has not been elucidated, it is clear that complexes containing CBP are active in reversing the inhibition, and of the components in these complexes, CBP is not affected while the 220-kd protein is. Furthermore, eIF-4A and eIF-4B, which appear to be the other two components of these active complexes, are also not altered (Duncan et al. 1983).

II. Reovirus

The poliovirus system is by far the best understood in terms of alteration of the host's ability to discriminate between capped and uncapped mRNA. Although less information is available, a similar transformation may occur during reovirus infection. Skup and Millward (1980) reported that normal L-cell lysates translate capped mRNA's well and uncapped mRNA's poorly, and that capped mRNA translation is sensitive to inhibition by cap analogues. In infected cell lysates, the reverse is true: uncapped mRNA is translated at high efficiency; capped mRNA is translated at reduced efficiency and is not sensitive to cap analogues. Furthermore, uncapped mRNA is more stable to nucleolytic degradation in infected cell lysates than in the uninfected control. Finally, these authors state that whereas reovirus mRNA's made early are capped, those which are made by the time the transition from a capped to uncapped translation occurs (20 h post-infection) are predominantly uncapped (Zarbl et al. 1980; Skup et al. 1981). Sonenberg et al. (1981b) further investigated this system using the IgM monoclonal antibody against CBP-related polypeptides (Sect. F.I). They found that capped reovirus mRNA translation was inhibited by antibody in uninfected cell lysates but that of uncapped EMC virus was not. In reovirus-infected cell lysates, no capped reovirus mRNA translation could be detected, but that of uncapped reovirus mRNA could. The latter was insensitive to inhibition by antibody. Thus, these results suggest that the reovirus-infected cell makes a transition over a 20-h period from cap-dependent to cap-independent translation.

A separate study by Detjen et al. (1982) has appeared which contradicts the findings discussed above. These authors used a different cell type, SC-1 cells, and basically found no changes in specificity for capped over uncapped mRNA translation occuring during reovirus infection. When they studied the L cell system used by Skup and Millward (1980), they observed a much greater shutoff of host translation than with the SC-1 cells, but still no differences in specificity. These authors contend that reovirus and host mRNA's compete for rate-limiting components of an unaltered protein synthetic machinery. The resolution of these differences may require careful comparison of the systems and methodologies used.

III. EMC Virus

Even though EMC virus and poliovirus are both picornaviruses, they appear to have different mechanisms for taking over host protein synthesis. Jen et al. (1980) compared the infection of HeLa cells by these viruses and found that, whereas host protein synthesis is shut down rapidly (2 h) and completely by polio, there is a more graduate transition from synthesis of host proteins to EMC virus proteins over approximately 6 h. Ribosomal salt washes from poliovirus-infected cells were inactive in permitting globin mRNA translation in a fractionated, initiation factor dependent system while those from EMC virus-infected cells was completely active. In a later study (Jen and Thach 1982), it was shown that the ability of HeLa cell lysates to translate host mRNA actually increased somewhat during EMC virus infection, peaked at 3.5 h, and then decreased thereafter. The cell-free system was fractionated into a 200,000 × g supernatant and pellet. The supernatant was found to increase in a factor which permitted translation of capped mRNA but had no effect on uncapped mRNA.

While this factor was not further characterized, the results are strikingly similar to those of Hansen et al. (1982), who showed that CBP (as determined by the cross-linking assay) was released from eIF-3 by poliovirus infection (Sect. G.I). Thus, while there are obvious differences between picornaviruses with respect to kinetics and severity of shutdown of host protein synthesis, there may be unifying themes in their mechanisms of action.

IV. Other Viruses

Several other host-virus systems have been studied with respect to changes in translational initiation. Detailed investigations comparable to those of poliovirus have not yet been conducted, but the evidence obtainted to date is suggestive of similar mechanisms.

One of these Semliki Forest virus. Steeg et al. (1981a,b) studied translation of EMC virus, globin, and both early and late mRNA of Semliki Forest virus in a reconstituted system. They found that the ribosomal wash fraction from infected cells was able to support translation of uncapped and late viral mRNA's but not capped or early viral RNA's. Late Semliki Forest virus mRNA is capped but not very susceptible to inhibition by cap analogues. Thus, it may be similar to alfalfa mosaic virus 4 and possibly contain very little secondary structure at the 5'-terminus (see Sect. E.III.3). As with poliovirus, additon of either eIF-4B or CBP restored the ability of infected cell ribosomal wash fractrions to stimulate capped mRNA translation.

Vaccinia virus mRNA was translated in a reticulocyte system (Lemieux and Beaud 1982). No competition between host and mRNA was detected as long as the RNA concentration was low. However, inhibition occured at high RNA concentrations. This could be reversed by initiation factors from infected but not uninfected Ehrlich ascites cells.

As a final example, frog virus 3 was translated in a number of cell-free systems (Raghow and Granoff 1983). Translation of early viral mRNA was efficient, but that of late mRNA was poor unless a ribosomal salt wash from infected cells was added. The block in late mRNA translation was shown to occur at the initiation step.

H. Cap-binding Protein and the Cytoskeleton

One of the most unexpected areas into which research on CBP has led is cell architecture and its relationship to protein synthesis. Three major lines of investigation have led to the postulate that mRNA and CBP are attached to a filament network in the cell. These will be the subject of this final section. A more complete review of this topic can be found in Nielsen et al. 1983).

I. Localization of mRNA on the Cytoskeleton

Penman (1982) has recently reviewed the pioneering work of his laboratory on the cytoskeletal network. To avoid duplication, only the major points will be recapitulated here. It is possible to separate the soluble elements of cultured cells from the structural proteins by treatment with non-ionic detergents, e.g., Triton X-100. This removes lipids and approximately two-thirds of the cellular proteins, leaving behind intact cytoskeletal components. Viewed in the electron microscope, extracted cells consist of a network of filaments of varying diameters, and surprisingly, all of the polyribosomes of the cell are bound to these filaments, usually at filament junctions (Lenk et al. 1977). Furthermore, disruption of the polysomes by inhibitors of protein synthesis allows the ribosomes to become soluble but leaves the mRNA bound to the cytoskeleton. The conclusion that it is the mRNA which is attached, not the ribosomes, is strengthened by the fact that only polysomes are found on the cytoskeleton; monosomes are present in the soluble phase of the cell. Studies with VSV show that viral mRNA initially associates with the cytoskeleton and forms polyribosomes but later dissociates and ceases to be translated (Cervera et al. 1981). This suggests that binding of the mRNA to the cytoskeleton may be obligatory for translation. Also, the VSV G-protein, which is translated only on ribosomes bound to the endoplasmic reticulum, is retained on cytoskeletons as well. This indicates that the cytoskeleton runs through the endoplasmic reticulum as well.

A different approach was used by Sharpe et al. (1980) to show a relationship between cytoskeleton and protein synthesis. They used diphtheria toxin, exotoxin A, and cycloheximide, all inhibitors of protein synthesis, in CV-1 cells and showed that the vimentin filament system (intermediate filaments) was disrupted. However, microtubules and microfilaments were not affected.

II. Antibodies to Cap-binding Protein-related Polypeptides

In a completely independent line of investigation, Zumbé et al. (1982) examined the cytoplasmic distribution of antigens reacting with the IgM antibody described above (Sect. F.I). Immunofluorescent staining of BHK cells unexpectedly revealed that the intermediate filament network of the cell bound the antibody. Control antisera against tubulin and actin revealed that the filaments were neither microtubules nor microfilaments. Double-labeling with anti-decamin antibodies, on the other hand, produced an identical pattern to that of the anti-CBP antibodies. Colchicine disrupted microtubules but not the intermediate filaments to which the anti-CBP antibodies were bound.

Further evidence for a cytoskeletal association came from studies using the Triton X-100 fractionation procedure described above (Sect. H.I). BHK cells were separated into soluble and cytoskeletal fractions and the protein constituents subjected to electrophoresis and immunoblotting. Immunoreactive polypeptides of 50 and 93 kd remained with the cytoskeleton, apparently the same as identified earlier in whole cells (Sect. F.I). The 50-kd polypeptide was shown to share tryptic peptides with CBP.

Taking these two observation together (Sect. I and II), one might postulate that mRNA is held to the cytoskeleton by association with CBP which itself is bound in some way to the intermediate filament network.

III. Viral Infection

As discussed above (Sect. G), infection of cells with several viruses appears not only to inactivate elements of the cap-recognition machinery but also alter their cytoplasmic distribution. If the mRNA is bound to the cytoskeleton via the cap-binding protein, one might reasonably expect that infection with poliovirus might release polysomes from the cytoskeleton by virtue of its inactivation of the cap-recognizing complex. In fact, Lenk and Penman (1979) showed that the cytoskeletal framework is reorganized following poliovirus infection. A central region of cytoplasm becomes modified and contains intermediate filaments in a unique pattern. Host mRNA is released from the cytoskeleton upon infection; polyribosomes carrying viral RNA appear to be on a rearranged type of cytoskeletal network.

Other viral systems show different effects. Vaccinia infection of Ehrlich ascites cells does not disrupt the association of host mRNA with cytoskeleton (Lemieux and Beaud 1982). However, a substantial portion of the viral mRNA in polysomes is not cytoskeleton-bound whereas nearly all of the host mRNA is. In adenovirus infection, some virus specific mRNA's associate more with the cytoskeleton than others (Mariman et al. 1982).

It is intriguing to speculate that the 220-kd polypeptide which is proteolytically cleaved during poliovirus infection (Sect. G.I) and which is an integral part of complexes containing CBP (Sect. E.II.5) may be the link between the cytoskeleton and CBP. This would correlate with the picornavirus-induced release of CBP into the cell supernatant (Sect. G, I and III) and release of polysomes from the intermediate filaments to the soluble phase (Lenk and Penman 1979).

J. Conclusion and Future Directions

As this review has hopefully demonstrated, research on the entry of mRNA into the cycle of protein synthesis is proceeding on many planes. There are now more questions which remain unanswered than are answered. However, some aspects of the mechanism appear to be established. Initial interaction of mRNA with the protein synthetic machinery is via the 5'-terminal cap, and this interaction imparts a five- to tenfold acceleration in the rate of binding to the 43S initiation complex. The only component with an active site for the cap appears to be the 24-kd CBP, but other factors such as eIF-4A and possibly eIF-4B form the subunits of a complex which carries out the process. ATP hydrolysis is required, and eIF-4A appears to be the factor responsible. This may be through an ATP-driven unwinding mechanism by which secondary structure is melted out of the 5'-terminus of the mRNA permitting the 43S ribosomal subunit of scan for the first AUG codon. A 220-kd polypeptide is also part of the complex, and this may serve as a link with eIF-3 or the intermediate filament network or both. This link appears to be the target whereby picornaviruses and perhaps other viruses alter the host protein synthetic apparatus to their advantage.

Questions remain at all levels of current research. Gradually the functional complexes of proteins will have to be built up from their components by defining specific interactions and partial reactions. As an example of such a partial reaction, one must consider the release of caps from the cap-binding protein: is the cap of each translated mRNA permanently bound to protein, or does CBP cycle with each round of translation? Is there an ATP-driven conformational change which releases the 5'-terminus from CBP?

The structural basis for differences in translational efficiency of different mRNA's remains a major unsolved problem. Why are some mRNA's more cap-dependent than others? What is a "cap-independent pathway" of initiation? Also, are there mechanisms of modulating the recognition of mRNA's which account for either overall regulation of the rate of protein synthesis or shifts in the spectrum of mRNA's translated?

Finally, and perhaps most intriguing, the chapter on the cell biology of protein synthesis is just beginning. Which elements are bound to the cytoskeletal framework, and what is their mode of attachment? Clearly protein synthesis can occur in cell-free extracts lacking a cytoskeleton. Yet the studies with poliovirus and VSV indicate that attachment of mRNA to the cytoskeleton is essential for translation. What properties result from its association with the filament system? Answers to these questions, which bridge the fields of biochemistry and cell biology, will require new research techniques as well as new ways of looking at protein synthesis.

K. 1984 Addendum

The foregoing review was completed in June, 1983. Between then and the present (October, 1984) a considerable amount of new information has been obtained on cap-binding protein and its interaction with mRNA. Published reports dealing with nearly all of the topics discussed above have appeared, and these will be briefly reviewed.

Three related review articles have appeared, dealing with the 5'-terminal cap structure of cellular and viral mRNA's (Moss 1984), the inhibition of cellular protein synthesis by polio and related viruses (Ehrenfeld 1984), and characteristics of eukaryotic mRNA CBP complex (Sonenberg et al. 1984).

Ishida et al. (1983) have demonstrated by the use of absorption, NMR and fluorescence spectroscopy that 7-methylguanine forms a 1:1 stacked structure with tryptophan. The authors speculate that the active site of CBP may contain a tryptophan residue as well as that of an acidic amino acid. Chemical synthesis of capped oligonucleotides has also recently been achieved. Yamaguchi et al. (1984) have extended their earlier work on synthesis of m^7GpppAm to include two capped oligonucleotides, m^7GpppAmpG and m^7GpppAmpGpU, which represent 5'-terminal structures from cytoplasmic polyhedrosis virus RNA.

The study described in Section C.V.3, the dependence of cap recognition on pH, has been published (Rhoads et al. 1983). In addition to the use of cap analogues as a probe of cap recognition, globin mRNA was treated with tobacco acid pyrophosphatase to remove the cap. Decapped mRNA exhibited a lesser dependence of translation on pH in the 6.5 - 8.0 range than its capped counterpart, again suggesting that the *enolate* form is preferentially recognized. A number of studies have also appeared describing a correlation between the physiological activation of cells and increases in intracellular pH (0.2 - 0.5 pH units). These include mitogenic stimulation of lymphocytes (Gerson et al. 1982), activation of respiration and motility of sea urchin sperm (Christen et al. 1982) and the transition from dormancy to development of brine shrimp embryos (Busa and Crowe 1983). While such correlations do not prove cause and effect, one cannot exclude the possibility that such changes in cell function result from a pH-dependent increase in protein synthesis initiation.

A description of the synthesis and use of the new affinity material for purification of CBP described in Section D.II.2 has now been published (Webb et al. 1984). The p-aminophenyl γ-ester of 7-methylguanosine 5'triphosphate coupled to Sepharose 4B is now commercially available from Pharmacia P-L Biochemicals, Inc. Using this, the preparation of CBP to approximately 90% purity from rabbit reticulocyte postribosomal supernatant can be achieved in 1 day.

A number of studies have been published dealing with the involvement of CBP with other initiation factors and with the ATP-dependent steps in initiation. Edery et al. (1983) investigated the 50-kd polypeptide present in high molecular weight complexes of CBP (termed eIF-4F; see Sect. E.II.5) and demonstrated that it reacted with antibodies to eIF-4A. By two-dimensional peptide mapping, the 50-kd protein and eIF-4A were also shown to be very similar. One peptide was clearly different, however, suggesting that a covalent modification of eIF-4A may be necessary in order to make it become part of eIF-4F. This 50-kd polypeptide was also shown to be the same species that cross-linked to caps in the presence of ATP and Mg^{2+} (Sect. E.I.3). Grifo et al. (1984) also studied eIF-4A and eIF-4F and demonstrated that they both possessed an RNA-dependent ATPase activity. This complements the earlier study in which these authors showed that ATP stimulates the CBP-dependent cross-linking of mRNA 5'-termini to eIF-4A

(Sect. E.I.3). They also presented evidence that ATP was not involved in phosphorylation but rather in hydrolysis to ADP and P_i. These studies suggest that the first of the two ATP-dependent steps in intiation is carried out by the combination of eIF-4A, eIF-4B and eIF-4F working in concert.

A potential mechanism by which these factors act, the ATP-dependent melting of secondary structure at the 5'-terminus of mRNA, is discussed above (Sect. E.III.3). Several studies supporting this hypothesis have appeared. Edery et al. (1984) translated capped mRNA's in a cell-free system containing elevated K^+ concentration, conditions expected to increase the secondary structure of mRNA. They demonstrated that translation was inhibited but could be reversed by eIF-4F. With a messenger containing little secondary structure at the 5'-terminus, RNA 4 of alfalfa mosaic viurs, such inhibition was not observed (Gelerke et al. 1983). A similar study by Sarkar et al. (1984) demonstrated that α globin mRNA was preferentially stimulated over β globin mRNA. The implication would be that α globin mRNA has more secondary structure at the 5'-terminus. Ray et al. (1984) provided more direct evidence for unwinding of mRNA secondary structure by eIF-4A. They subjected 5'-end labeled reovirus mRNA to mild nuclease attack as a probe for secondary structure. In the presence of ATP and eIF-4A, increased cleavage was observed, which was interpreted as partial melting of the mRNA secondary structure. This effect was also observed with eIF-4F. Equivalent unwinding was observed with only 1/25th the amount of eIF-4A (contained within the eIF-4F complex) as with free eIF-4A, suggesting that the other components of eIF-4F (CBP, P220, and possibly eIF-4B) greatly enhance the ATP-dependent unwinding activity of eIF-4A. Evidence against this hypothesis, however, was presented by Tahara et al. (1983). A cell-free system was depleted of ATP with hexokinase and glucose. Under these conditions, initiation complex formation was completely dependent on added ATP. When mRNA substituted with inosine to reduce secondary structure was used, initiation complex formation was still ATP-dependent. The same result was obtained using a system in which cross-linking of mRNA 5'-termini to eIF-4A and eIF-4B was followed. The authors concluded that if an ATP-dependent denaturation occurs, it must be after the mRNA has become attached to the 40S ribosomal subunit. A possible resolution of this discrepancy, recalling that two ATP-dependent steps during initiation have been proposed (Sect. E.I.3), is that mRNA binding to CBP (or an immediately subsequent event) requires ATP, and that this is not affected by changes in mRNA secondary structure. Then there is an ATP-dependent unwinding, driven by eIF-4A and facilitated by components in eIF-4F, which accomplishes the scanning by 40S ribosomal subunits for the first AUG. Unfortunately, however, as Grifo et al. (1984) point out, there have been no reports of a 48S initiation complex which contains eIF-4A, -4B or -4F. Thus, the overall mechansim, while clarified somewhat by these recent contributions, has not yet been fully elucidated.

Section F deals with the subject of antibodies related to CBP or associated polypeptides. As mentioned above, the three antibodies described functioned in ways expected of an anti-CBP antibody (inhibition of capped mRNA translation, inhibition of cross-linking between the 5'-termini of mRNA and various polypeptides) but did not react with CBP itself, as judged by the immunoblotting technique. Our own laboratory has recently prepared a polyclonal antibody in goats using affinity-purified CBP as antigen. Briefly, the following results have been obtained: (1) On immunoblotting with crude initiation factors (high salt ribosomal extract), the 24-kd CBP predominantly reacted. Some reaction with polypeptides of higher molecular weight was also observed. Pretreatment of the antibody with highly purified CBP (affinity chromatography followed by high performance liquid chromatography) eliminated the reaction with the 24-kd species but failed to affect the reaction with larger polypeptides. Thus, the latter represents nonspecific binding of some component of the immunoblotting system. These results suggest that, in contrast to the results cited above, there is only one species of CBP in crude initiation factors. (2) The anti-CBP antibodies inhibited the translation of a typical capped mRNA (globin) but failed to inhibit that of an uncapped mRNA (STNV). (3) Whole reticulocyte lysate was fractioned on a sucrose gradient and fractions probed with the antibody. CBP was found in the supernatant fraction and the ribosomal subunits (in a ratio of approximately 2:1, respectively) but not in 80S ribosomes or polysomes. (4) Finally, CBP was purified by conventional means (Sect. D.II.1) and the presence of polypeptides related to CBP probed by immunoblotting. No evidence for high molecular weight precursors for CBP was found, suggesting that the 24-kd species is not proteolytically produced during purification.

The inactivation of cellular protein synthesis by viruses is discussed in Section G. Etchison et al. (1984) have recently established assays to independently measure the activities of eIF-3 and the CBP-complex after infection of cells with poliovirus. Since CBP is associated with eIF-3 during purification, it was not clear which of these components was inactivated by poliovirus. The results indicated that only CBP complex was affected and that eIF-3 retained full activity in poliovirus-infected cells. Further evidence has been obtained that Semliki Forest virus interferes with the function of CBP (see Sect. G.IV). Van Steeg et al. (1984) demonstrated that translation of host mRNA in in vitro systems from virus-infected neuroblastoma cells could be restored with preparations containing CBP. Unlike poliovirus, the capsid protein of Semliki Forest virus appeared to specifically prevent the binding of host mRNA to ribosomes. A virus not mentioned originally in Section G is vesicular stomatitis virus (VSV). Thomas and Wagner (1983) have studied VSV-infected L cell lysates and found that translation was depressed relative to uninfected controls. The cell-free system was judged to be a good model of the situation in intact cells since the degree of inhibition was similar to that in infected cells. By fractionating lysates and reconstituting components from control and virus-infected cells, the defect was shown to be in the 0 - 40% ammonium sulfate fraction of ribosomal salt wash. The only initiation factors traditionally assigned to this fraction are eIF-3, eIF-4B and CBP. Thus, alternations of the cap-recognizing machinery similar to those observed with other viruses may occur with VSV as well.

Three additional studies dealing with the relationship of the cytoskeleton to the mRNA recognition step in initiation have been published. Jeffery (1982) used two techniques to examine the subcellular location of mRNA in ascidian follicle cells, extraction with Triton X-100 and in situ hybridization. He determined that the cytoskeleton contained 60 - 70% of the total mRNA of the cell, but over 90% of the newly synthesized mRNA. Howe and Hershey (1984) also used detergent to fractionate HeLa cells but monitored the distribution of initiation factors. They found that eIF-2, eIF-3, eIF-4A and eIF-4B were all enriched in the cytoskeleton. Release of both ribosomes and initiation factors into the soluble fraction could be effected by heat shock, fluoride, pactamycin or cytochalasin. Finally, Patzelt et al. (1983) have explored the subcellular distribution of CBP using a novel photoaffinity probe, γ-[^{32}P]-[4-(bezoylphenyl)methylamido]-7-methylguanosine-5'-triphosphate. Whole cell homogenates of HeLa cells were labeled and fractionated. In addition to the cytoplasmic 24-kd CBP, polypeptides of 37, 80, 89, and 120 kd were labeled. Some were found to be associated with the nucleus. Such studies may lead to the discovery of a nuclear CBP whose function is to recognize mRNA by virtue of its unique 5'-terminal structure and participate in some aspects of mRNA processing or export to the cytoplasm.

Acknowledgements. I wish to thank Ms. Nancy Webb and Ginger Hodges for invaluable assistance with this manuscript and Dr. Robert Thach for making manuscripts available to me before publications.

References

Abraham, K.A., Lillehaug, J.R.: Enzymatic cleavage of m^{7}GDP from eucaryotic mRNA. FEBS Lett. 71, 49-52 (1976)

Adams, J.M., Cory, S.: Modified nucleosides and bizzare 5'-termini in mouse myeloma mRNA. Nature (London) 255, 28-33 (1975)

Adams, B.L., Morgan, M., Muthukrishnan, S., Hecht, S.M., Shatkin, A.J.: The effect of "cap" analogs on reovirus mRNA binding to wheat germ ribosomes. J. Biol. Chem. 253, 2589-2595 (1978)

Asselsbergs, A.M., Peters, W., Van Venrooij, W.J., Bloemendal, H.: Diminished sensitivity of re-initiation of translation to inhibition by cap analogues in reticulocyte lysates. Eur. J. Biochem. 88, 483-488 (1978a)

Asselsberg, A.M., Peters, W., Van Venrooij, W.J., Bloemendal, H.: Inhibition of translation of lens mRNAs in a messenger dependent reticulocyte lysate by cap analogues. Biochim. Biophys. Acta 520, 577-587 (1978b)

Baglioni, C., Simili, M., Shafritz, D.A.: Initiation activity of EMC virus RNA, binding to initiation factor eIF-4B and shut-off of host cell protein synthesis. Nature (London) 275, 240-243 (1978)

Banerjee, A.K.: 5'-Terminal cap structure in eucaryotic messenger ribonucleic acids. Microbiol. Rev. 44, 175-205 (1980)

Baralle, F.E., Brownlee, G.G.: AUG is the only recognisable signal sequence in the 5' non-coding regions of eukaryotic mRNA. Nature (London) 274, 84-87 (1978)

Benne, R., Hershey, J.W.B.: The mechanism of action of protein synthesis initiation factors from rabbit reticulocytes. J. Biol. Chem. 253, 3078-3087 (1978)

Bergmann, J.E., Trachsel, H., Sonenberg, N., Shatkin, A.J., Lodish, H.F.: Characterization of rabbit reticulocyte factor(s) that stimulates the translation of mRNAs lacking 5'-terminal 7-methylguanosine. J. Biol. Chem. 254, 1440-1443 (1979)

Blaas, D., Patzelt, E., Kuechler, E.: Identification of the cap binding protein of influenza virus. Nucleic Acids Res. 10, 4803-4812 (1982)

Both, G.W., Banerjee, A.K., Shatkin, A.J.: Methylation-dependent translation of viral messenger RNAs in vitro. Proc. Natl. Acad. Sci. USA 72, 1189-1193 (1975a)

Both, G.W., Furuichi, Y., Muthukrishnan, S., Shatkin, A.J.: Ribosome binding to reovirus mRNA in protein synthesis requires 5'-terminal 7-methylguanosine Cell 6, 185-195 (1975b)

Both, G.W., Furuichi, Y., Muthukrishnan, S., Shatkin, A.J.: Effect of 5'-terminal structure and base composition on polyribonucleotide binding to ribosomes. J. Mol. Biol. 104, 637-658 (1976)

Brether, H.-J., Malek, L.T., Hellmann, G.M., Rhoads, R.E.: Determination of 7-methylguanosine-containing 5'-termini of messenger ribonucleic acid by $NaB[^3H]_4$ labeling and high performance liquid anion exchange chromatography. Anal. Biochem. 98, 102-111 (1979)

Busa, W.B., Crowe, J.H.: Intracellular pH regulates transition between dormancy and development of brine shrimp *(Artemia salina)* embryos. Science 221, 366-368 (1983)

Canaani, D., Revel, M., Groner, Y.: Translational discrimination of "capped" and "non-capped" mRNAs: Inhibition by a series of chemical analogs of m^7GpppX. FEBS Lett. 64, 326-331 (1976)

Cervera, M., Dreyfuss, G., Penman, S.: Messenger RNA is translated when associated with the cytoskeletal framework in normal and VSV-infected HeLa cells. Cell 23, 113-120 (1981)

Checkley, J.W., Cooley, L., Ravel, J.M.: Characterization of initiation factor eIF-3 from wheat germ. J. Biol. Chem. 256, 1582-1586 (1981)

Christen, R., Schackmann, R.W., Shapiro, B.M.: Elevation of the intracellular pH activates respiration and motility of sperm of the sea urchin, *Strongylocentrotus purpuratus*. J. Biol. Chem. 257, 14881-14890 (1982)

Chu, L.-Y., Rhoads, R.E.: Translational recognition of the 5'-terminal 7-methylguanosine of globin mRNA as a function of ionic strength. Biochemistry 17, 2540-2455 (1978)

Chu, L.-Y., Rhoads, R.E.: Inhibition of cell-free messenger ribonucleic acid translation by 7-methylguanosine 5'-triphosphate: Effect of messenger ribonucleic acid concentration. Biochemistry 19, 184-191 (1980)

Chu, L.-Y., Lockard, R.E., Rajbhandary, U.L., Rhoads, R.E.: Paradoxical observations on the 5'-terminus of ovalbumin messenger ribonucleic acid. J. Biol. Chem. 253, 5228-5231 (1978)

Daniels McQueen, S., Detjen, B.M., Grifo, J.A., Merrick, W.C., Thach, R.E.: Unusual requirements for optimum translation of polio viral RNA in vitro. J. Biol. Chem. 258, 7195-7199 (1983)

Darzynkiewicz, E., Antosiewicz, J., Ekiel, I., Morgan, M.A., Tahara, S.M., Shatkin, A.J.: Methyl esterification of $m^7G^{5'}$p reversibly blocks its activity as an analog of eukaryotic mRNA 5'-caps. J. Mol. Biol. 153, 451-458 (1981)

Detjen, B.M., Walden, W.E., Thach, R.E.: Translational specificity in reovirus-infected mouse fibroblasts. J. Biol. Chem. 257, 9855-9860 (1982)

Duncan, R., Etchison, D., Hershey, J.W.B.: Protein synthesis eukaryotic initiation factors eIF4A and eIF4B are not altered by poliovirus infection of HeLa cells. J. Biol. Chem. 258, 7236-7239 (1983)

Edery, I., Humbelin, M., Darveau, A., Lee, K.A.W., Milburn, S., Hershey, J.W.B., Trachsel, H., Sonenberg, N.: Involvement of eukaryotic initiation factor 4A in the cap recognition process. J. Biol. Chem. 258, 11398-11403 (1983)

Edery, I., Lee, K.A.W., Sonenberg, N.: Functional characteristics of eukaryotic mRNA cap binding protein complex: Effects on translation of capped and naturally uncapped RNAs. Biochemistry 23, 2456-2462 (1984)

Efstratiadis, A., Vournakis, J.N., Donis-Keller, H., Chaconas, G., Dougall, D.K., Kafatos, F.C.: End labeling of enzymatically decapped mRNA. Nucleic Acids Res. 4, 4165-4174 (1977)

Ehrenfeld, E.: Poliovirus-induced inhibition of host-cell protein synthesis. Cell 28, 435-436 (1982)

Ehrenfeld, E.: Picornavirus inhibition of host cell protein synthesis. In: Comprehensive Virology (eds. H. Fraenkel-Conrat, R.R. Wagner). New York: Plenum Press 1984

Etchison, D., Milburn, S.C., Edery, I., Sonenberg, N., Hershey, J.W.B.: Inhibition of HeLa cell protein synthesis following poliovirus infection correlates with the proteolysis of a 220,000-dalton polypeptide associated with eukaryotic initiation factor 3 and a cap binding protein complex. J. Biol. Chem. 257, 14806-14810 (1982)

Etchinson, D., Hansen, J., Ehrenfeld, E., Edery, I., Sonenberg, N., Milburn, S., Hershey, J.W.B.: Demonstration in vitro that eukaryotic initiation factor 3 is active but cap-binding protein complex is inactive in poliovirus-infected HeLa cells. J. Virol. 51, 832-837 (1984)

Filipowicz, W., Furuichi, Y., Sierra, J.M., Muthkrishnan, S., Shatkin, A.J., Ochoa, S.: A protein binding the methylated 5'-terminal sequence, m^7GpppN, of eukaryotic messenger RNA. Proc. Natl. Acad. Sci. USA 73, 1559-1563 (1976)

Furuichi, Y., Miura, K.: A blocked structure at the 5'-terminus of mRNA from cytoplasmic polyhedrosis virus. Nature (London) 253, 374-375 (1975)

Furuichi, Y., Muthukrishnan, S., Shatkin, A.J.: 5'-Terminal m^7G(5')ppp(5')G^mp in vivo: Identification in reovirus genome RNA. Proc. Natl. Acad. Sci. USA 72, 742-745 (1975)

Furuichi, Y., Lafiandra, A., Shatkin, A.J.: 5'-Terminal structure and mRNA stability. Nature (London) 266, 235-239 (1977)

Furuichi, Y., Morgan, M.A., Shatkin, A.J.: Synthesis and translation of mRNA containing 5'-terminal 7-ethylguanosine cap. J. Biol. Chem. 254, 6732-6738 (1979)

Gehrke, L., Auron, P.E., Quigley, G.J., Rich, A., Sonenberg, N.: 5'-Conformation of capped alfalfa mosaic virus ribonucleic acid 4 may reflect its independence of the cap structure or of cap-binding protein for efficient translation. Biochemistry 22, 5157-5164 (1983)

Gerson, D.F., Kiefer, H., Eufe, W.: Intracellular pH of mitogen-stimulated lymphocytes. Science 216, 1009-1010 (1982)

Grifo, J.A., Tahara, S.M., Leis, J.P., Morgan, M.A., Shatkin, A.J., Merrick, W.C.: Characterization of eukaryotic initiation factor 4A, a protein involved in ATP-dependent binding of globin mRNA. J. Biol. Chem. 257, 5246-5252 (1982)

Grifo, J.A., Tahara, S.M., Morgan, M.A., Shatkin, A.J., Merrick, W.C.: New initiation factor activity required for globin mRNA translation. J. Biol. Chem. 258, 5804-5810 (1983)

Grifo, J.A., Abramson, R.D., Satler, C.A., Merrick, W.C.: RNA stimulated ATPase activity of eukaryotic initiation factors. J. Biol. Chem. 259, 8648-8654 (1984)

Hansen, J.L., Ehrenfeld, E.: Presence of the cap-binding protein in initiation factor preparations from poliovirus-infected HeLa cells. J. Virol. 38, 438-445 (1981)

Hansen, J.L., Etchison, D.O., Hershey, J.W.B., Ehrenfeld, E.: Localization of cap-binding protein in subcellular fraction of HeLa cells. Mol. Cell. Biol. 2, 1639-1643 (1982a)

Hansen, J.L., Etchison, D., Hershey, J.W.B., Ehrenfeld, E.: Association of cap-binding protein with eukaryotic initiation factor 3 in initiation factor preparations from uninfected and poliovirus-infected HeLa cells. J. Virol. 42, 200-207 (1982b)

Hari, V.: The RNA of tobacco etch virus: Further characterization and detection of protein linked to RNA. Virology 112, 391-399 (1981)

Hattori, M., Miura, K., Yamaguchi, K., Ohtani, S., Hata, T.: Interaction between bases involved in the 5'-terminal cap structure of eukaryotic mRNA. Nucleic Acids Res. Spec. Publ. No. 5 (1978)

Hecht, S.M., Adams, B.L., Kozarich, J.W.: Chemical transformations of 7,9-disubstituted purines and related heterocycles. Selective reduction of imines and immonium salts. J. Org. Chem. 41, 2305-2311 (1976)

Held, W.A., West, K., Gallagher, J.F.: Importance of initiation factor preparations in the translation of reovirus and globin mRNAs lacking a 5'-terminal 7-methylguanosine. J. Biol. Chem. 252, 8489-8497 (1977)

Helentjaris, T., Ehrenfeld, E.: Control of protein synthesis in extracts from poliovirus-infected cells. I. mRNA discrimination by crude initiation factors. J. Virol. 26, 510-521 (1978)

Helentjaris, T., Ehrenfeld, E., Brown-Luedi, M.L., Hershey, J.W.B.: Alterations in initiation factor activity from poliovirus-infected HeLa cells. J. Biol. Chem. 254, 10973-10978 (1979)

Hellmann, G.M., Shaw, J.G., Lesnaw, J.A., Chu, L.-Y., Pirone, T.P., Rhoads, R.E.: Cell-free translation of tobacco vein mottling virus RNA. Virology 106, 207-216 (1980)

Hellmann, G.M., Chu, L.-Y., Rhoads, R.E.: A polypeptide which reverses cap analogue inhibition of cell-free protein synthesis. J. Biol. Chem. 257, 4056-4062 (1982)

Hendler, S., Furer, E., Srinivasan, P.R.: Synthesis and chemical properties of monomers and polymers containing 7-methylguanine and an investigation of their substrate or template properties for bacterial DNA or RNA polymerase. Biochemistry 9, 4141-4153 (1970)

Hickey, E.D., Weber, L.A., Baglioni, C.: Inhibition of initiation of protein synthesis by 7-methylguanosine-5'-monophosphate. Proc. Natl. Acad. Sci. USA 73, 19-23 (1976)

Hickey, E.D., Weber, L.A., Baglioni, C., Kim, C.H., Sarma, R.H.: A relation between inhibition of protein synthesis and conformation of 5'-phosphorylated 7-methylguanosine derivatives. J. Mol. Biol. 109, 173-183 (1977)

Howe, J.G., Hershey, J.W.B.: Translational initiation factor and ribosome association with the cytoskeletal framework fraction from HeLa cells. Cell 37, 85-93 (1984)

Ishida, T., Katsuta, M., Inoue, M., Yamagata, Y., Tomita, K.: The stacking interactions in 7-methylguanine-tryptophan systems, a model study for the interaction between the 'cap' structure of mRNA and its binding protein. Biochem. Biophys. Res. Commun. 115, 849-854 (1983)

Jagus, R., Anderson, W.F., Safer, B.: The regulation of initiation of mammalian protein synthesis. In: Progress in Nucleic Acid Research and Molecular Biology (eds. W.E. Cohn, E. Volkin), Vol. XXV, pp. 127-185. London-New York: Academic Press 1981

Jeffery, W.R.: Messenger RNA in the cytoskeletal framework: Analysis by in situ hybridization. J. Cell Biol. 95, 1-7

Jen, G., Thach, R.E.: Inhibition of host translation in encephalomyocarditis virus-infected L cells: a novel mechanism. J. Virol. 43, 250-261 (1982)

Jen, G., Detjen, B.M., Thach, R.E.: Shutoff of HeLa cell protein synthesis by encephalomyocarditis virus and poliovirus: a comparative study. J. Virol. 35, 150-156 (1980)

Kabat, D., Chappell, M.R.: Competition between globin messenger ribonucleic acids for a discriminating initiation factor. J. Biol. Chem. 252, 2684-2690 (1977)

Kaempffer, R., Rosen, H., Israeli, R.: Translation control: Recognition of the methylated 5' end and an internal sequence in eukaryotic mRNA by the initiation factor that binds met-tRNA$_f^{met}$. Proc. Natl. Acad. Sci. USA 75, 650-654 (1978)

Kemper, B., Stolarsky, L.: Dependence on potassium concentration of the inhibition of the translation of messenger ribonucleic acid by 7-methylguanosine 5'-phosphate. Biochemistry 16, 5676-5680 (1977)
Kim, C.H., Sarma, R.H.: Spatial configuration of mRNA 5'-terminus. Nature (London) 270, 223-227 (1977)
Koch, G., Bilello, J.A., Kruppa, J., Koch, F., Oppermann, H.: Amplification of translational control by membrane-mediated events: a pleiotropic effect on cellular and viral gene expression. Ann. N.Y. Acad. Sci. 339, 280-307 (1980)
Kozak, M.: How do eucaryotic ribosomes select initiation regions in messenger RNA? Cell 15, 1109-1123 (1978)
Kozak, M.: Role of ATP in binding and migration of 40S ribosomal subunits. Cell 22, 459-467 (1980a)
Kozak, M.: Influence of mRNA secondary structure on binding and migration of 40S ribosomal subunits. Cell 19, 79-90 (1980b)
Kozak, M., Shatkin, A.J.: Identification of features in 5'-terminal fragments from reovirus mRNA which are important for ribosome binding. Cell 13, 201-212 (1978)
Kozarich, J.W., Deegan, J.L.: 7-Methylguanosine-dependent inhibition of globin mRNA translation by methylglyoxal. J. Biol. Chem. 254, 9345-9348 (1979)
Kroath, H., Shatkin, A.J.: mRNA 5'-cap binding activity in purified influenza virus detected by simple, rapid assay. J. Virol. 41, 1105-1108 (1982)
Lee, K.A.W., Sonenberg, N.: Inactivation of cap-binding proteins accompanies the shut-off of host protein synthesis by poliovirus. Proc. Natl. Acad. Sci. USA 79, 3447-3451 (1982)
Lee, K.A.W., Guertin, D. Sonenberg, N.: mRNA secondary structure as a determinant in cap recognition and initiation complex formation. J. Biol. Chem. 258, 707-710 (1983)
Lemieux, R., Beaud, G.: Expression of vaccinia virus early mRNA in Ehrlich ascites tumor cells. 2. Part of the polysomes at an early stage of virus infection are not bound to the cytoskeleton. Eur. J. Biochem. 129, 273-279 (1982)
Lenk, R., Penman, S.: The cytoskeletal framework and poliovirus metabolism. Cell 16, 289-301 (1979)
Lenk, R., Ransom, L., Kaufmann, Y., Penman, S.: A cytoskeletal structure with associated polyribosomes obtained from HeLa cells. Cell 10, 67-78 (1977)
Lockard, R.E., Rajbhandary, U.L.: Nucleotide sequences at the 5'-termini of rabbit α and β globin mRNA. Cell 9, 747-760 (1976)
Lodish, H.F.: Translational control of protein synthesis. In: Annual Review of Biochemistry (eds. E.E. Snell, P.D. Boyer, A. Meister, C.C. Richardson), Vol. 45. Palo Alto. California: Annual Review Inc. 1976
Lodish, H.F., Rose, J.K.: Relative importance of 7-methylguanosine in ribosome binding and translation of vesicular stomatitis virus mRNA in wheat germ and reticulocyte cell-free systems. J. Biol. Chem. 252, 1181-1188 (1977)
Malek, L.T., Eschenfeldt, W.H., Munns, T.W., Rhoads, R.E.: Heterogeneity of the 5'-terminus of ovalbumin messenger ribonucleic acid. Nucleic Acids Res. 9, 1657-1673 (1981)
Marcus, A.: Tobacco mosaic virus ribonucleic acid-dependent amino acid incorporation in a wheat germ embryo system in vitro. Analysis of the rate-limiting reaction. J. Biol. Chem. 245, 955-961 (1970a)
Marcus, A.: Tobacco mosaic virus ribonucleic acid dependent amino acid incorporation in a wheat germ embryo system in vitro. Formation of a ribosome-messenger "initiation" complex. J. Biol. Chem. 245, 962-966 (1970b)
Mariman, E., Hagebols, A.-M., Van Venrooij, W.: On the localization and transport of specific adenoviral mRNA-sequences in the late infected HeLa cell. Nucleic Acids Res. 10, 6131-6145 (1982)

Mayo, M.A., Barker, H., Harrison, B.D.: Evidence for a protein covalently linked to tobacco ringspot virus RNA. J. Gen. Virol. 43, 735-740 (1979)

Morgan, M.A., Shatkin, A.J.: Initiation of reovirus transcription by inosine 5'-triphosphate and properties of 7-methylinosine-capped, inosine-substituted messenger RNA. Biochemistry 19, 5960-5966 (1980)

Moss, B.: 5'-Terminal cap structures of eukaryotic and viral mRNAs. In: Processing of RNA (ed. D. Apirion), pp. 119-127. Florida: CRC Press 1984

Muthukrishnan, S., Both, G.W., Furuichi, Y., Shatkin, A.J.: 5'-Terminal 7-methylguanosine in eukaryotic mRNA is required for translation. Nature (London) 255, 33-37 (1975)

Muthukrishnan, S., Morgan, M., Banerjee, A.K., Shatkin, A.J.: Influence of 5'-terminal m^7G and 2'-O-methylated residues on messenger RNA binding to ribosomes. Biochemistry 15, 5761-5768 (1976)

Muthukrishnan, S., Moss, B., Cooper, J.A., Maxwell, E.S.: Influence of 5'-terminal cap structure on the initiation of translation of vaccinia virus mRNA. J. Biol. Chem. 253, 1710-1715 (1978)

Nielsen, P., Goelz, S., Trachsel, H.: The role of the cytoskeleton in eukaryotic protein synthesis. Cell Biol. Int. Rep. 7, 245-254 (1983)

Nishimura, Y., Yamamoto, T., Tsuboi, M.: The nature of the fluorescence spectra of some minor bases obtained from tRNA and mRNA. Nucleic Acids. Res. Spec. Publ. No. 3, s85-s88 (1977)

Nishimura, Y., Takahashi, S., Yamamoto, T., Tsuboi, M., Hattori, M., Miura, K., Yamaguchi, K., Ohtani, S., Hata, T.: On the base-stacking in the 5'-terminal cap structure of mRNA: a fluorescence study. Nucleic Acids. Res. 8, 1107-1119 (1980)

Padilla, M., Canaani, D., Groner, Y., Weinstein, J.A., Bar-Joseph, M., Merrick, W., Shafritz, D.A.: Initiation factor eIF-4B (IF-M_3)-dependent recognition and translation of cappes versus uncapped eukaryptic mRNAs. J. Biol. Chem. 253, 5939-5945 (1978)

Parets-Soler, A., Reibel, L., Shapira, G.: Differential stimulation of α and β-globin mRNA translation by M_r 50,000 and 28,000 polypeptide containing fractions isolated from reticulocyte polysomes. FEBS Lett. 136, 259-264 (1981)

Patzelt, E., Blaas, D., Kuechler, E.: Cap binding proteins associated with the nucleus. Nucleic Acids Res. 11, 5821-5835 (1983)

Pelham, H.R.B., Jackson, R.J.: An efficient mRNA-dependent translation system from reticulocyte lysates. Eur. J. Biochem. 67, 247-256 (1976)

Penman, S.: Protein synthesis for cell architecture. In: Interaction of Translational and Transcriptional Controls in the Regulation of Gene Expression (eds. M. Grunberg-Manago, B. Safer), Vol. XXIV, pp. 243-264. Amsterdam-New York: Elsevier Science Publishing Company 1982

Pochon, F., Pascal, Y., Pitha, P., Michelson, A.M.: Photochimie des polynucleotides. IV. Photochimie de quelques nucleosides puriques methyles. Biochim. Biophys. Acta 213, 273-281 (1970)

Raghow, R., Granoff, A.: Cell-free translation of frog virus 3 messenger RNAs. J. Biol. Chem. 258, 571-578 (1983)

Ray, B.K., Brendler, T.G., Adya, S., Daniels-McQueen, S., Miller, J.K., Hershey, J.W.B., Grifo, J.A., Merrick, W.C., Thach, R.E.: Role of mRNA competition in regulating translation: Further characterization of mRNA discriminatory initiation factors. Proc. Natl. Acad. Sci. USA 80, 663-667 (1983)

Ray, B.K., Cladaras, M.H., Grifo, J.A., Abramson, R.D., Merrick, W.C., Thach, R.E.: How mRNA structure is unwound after recognition by translation initiation factors. Personal communication

Rhoads, R.E., Hellmann, G.M., Remy, P., Ebel, J.-P.: Translational recognition of messenger RNA caps as a function of pH. Biochemistry 22, 6084-6088 (1983)

Roman, R., Brooker, J.D., Seal, S.N., Markus, A.: Inhibition of the transition of a 40S ribosome-Met-tRNA$_i^{Met}$ complex to an 80S ribosome-Met-tRNA$_i^{Met}$ complex by 7-methylguanosine-5'-phosphate. Nature (London) 260, 359-360 (1976)
Rose, J.K., Lodish, H.F.: Translation in vitro of vesicular stomatitis virus mRNA lacking 5'-terminal 7-methylguanosine. Nature (London) 262, 32-37 (1976)
Rose, J.K., Trachsel, H., Leong, K., Baltimore, D.: Inhibition of translation by poliovirus: inactivation of a specific initiation factor. Proc. Natl. Acad. Sci. USA 75, 2732-2736 (1978)
Rosenberg, M., Paterson, B.M.: Efficient cap-dependent translation of polycistronic prokaryotic mRNAs is restricted to the first gene in the operon. Nature (London) 279, 696-701 (1979)
Rottman, F.M.: Methylation and polyadenylation of heterogeneous nuclear and messenger RNA. In: Biochemistry of Nucleic Acids II (ed. B.F.C. Clark), Vol. XVII, pp. 45-73. Baltimore, Maryland: University Park Press 1978
Rottman, E.M., Shatkin, A.J., Perry, R.P.: Sequences containing methylated nucleotides at the 5'-termini of messenger RNA's: Possible implications for processing. Cell 3, 197-199 (1974)
Rupprecht, K.M., Sonenberg, N., Shatkin, A.J., Hecht, S.M.: Design and preparations of affinity columns for the purification of eukaryotic messenger ribonucleic acid cap binding protein. Biochemistry 20, 6570-6577 (1981)
Sarkar, G., Edery, I., Gallo, R., Sonenberg, N.: Preferential stimulation of rabbit α globin mRNA translation by a cap binding protein complex. Biochim. Biophys. Acta, in press (1984)
Sasavage, N.L., Friderici, K., Rottman, F.M.: Specific inhibition of capped mRNA translation in vitro by $m^7G^{5'}pppp^{5'}G$ and $m^7G^{5'}pppp^{5'}m^7G$. Nucleic Acids Res. 6, 3613-3624 (1979)
Seal, S.N., Schmidt, A., Tomaszewski, M., Marcus, A.: Inhibition of noncapped mRNA translation by the cap analogue, 7-methylguanosine-5'-phosphate. Biochem. Biophys. Res. Commun. 82, 553-559 (1978)
Shafritz, D.A., Weinstein, J.A., Safer, B., Merrick, W.C., Weber, L.A., Hickey, E.D., Baglioni, C.: Evidence for role of $m^7G^{5'}$-phosphate group in recognition of eukaryotic mRNA by initiation factor IF-M_3. Nature (London) 261, 291-294 (1976)
Shapiro, R.: Chemistry of guanine and its biologically significant derivatives. Prog. Nucleic Acid Res. Mol. Biol. 8, 73-112 (1968)
Sharma, O.K., Hruby, D.E., Beezley, D.N.: Inhibition of ovalbumin mRNA translation by 7-methylguanosine-5'-phosphate. Biochem. Biophys. Res. Commun. 72, 1392-1398 (1976)
Sharpe, A.H., Chen, L.B., Murphy, J.R., Fields, B.N.: Specific disruption of vimentin filament organization in monkey kidney CV-1 cells by diphtheria toxin, exotoxin A, and cycloheximide. Proc. Natl. Acad. Sci. USA 77, 7267-7271 (1980)
Shatkin, A.J.: Capping of eukaryotic mRNAs. Cell 9, 645-653 (1976)
Shatkin, A.J., Darzynkiewicz, E., Furuichi, Y., Kroath, H., Morgan, M.A., Tahara, S.M., Yamakawa, M.: 5'-Terminal caps, cap-binding proteins, and eukaryotic mRNA function. Biochem. Soc. Symp. 47, 129-143 (1981)
Shibata, H., Ro-Choi, T.S., Reddy, R., Choi, Y.C., Henning, D., Busch, H.: The primary nucleotide sequence of nuclear U-2 ribonucleic acid. J. Biol. Chem. 250, 3909-3920 (1975)
Shimotohno, K., Kodama, Y., Hashimoto, J., Miura, K.: Importance of 5'-terminal blocking structure to stabilize mRNA in eukaryotic protein synthesis. Proc. Natl. Acad. Sci. USA 74, 2734-2738 (1977)
Skup, D., Millward, S.: Reovirus-induced modification of cap-dependent translation in infected cells. Proc. Natl. Acad. Sci. USA 77, 152-156 (1980)
Skup, D., Zarbl, H., Millward, S.: Regulation of translation in L-cells infected with reovirus. J. Mol. Biol. 151, 35-55 (1981)

Smith, R.E., Clark, J.M.: Effect of capping upon the mRNA properties of satellite tobacco necrosis virus ribonucleic acid. Biochemistry 18, 1366-1371 (1979)

Sonenberg, N.: ATP/Mg^{+++}-dependent cross-linking of cap binding proteins to the 5' end of eukaryotic mRNA. Nucleic Acids Res. 9, 1643-1656 (1981)

Sonenberg, N., Shatkin, A.J.: Reovirus mRNA can be covalently crosslinked via the 5'-cap to proteins in initiation complexes. Proc. Natl. Acad. Sci. USA 74, 4288-4292 (1977)

Sonenberg, N., Shatkin, A.J.: Nonspecific effect of m^7GMP on protein-RNA interactions. J. Biol. Chem. 253, 6630-6632 (1978)

Sonenberg, N., Trachsel, H.: Probing the function of the eukaryotic 5'-cap structure using monoclonal antibodies to cap-binding proteins. In: Current Topics in Cellular Regulation (eds. B.L. Horecker, E.R. Stadtman), Vol. XXI, pp. 65-88. London-New York: Academic Press 1982

Sonenberg, N., Morgan, M.A., Merrick, W.C., Shatkin, A.J.: A polypeptide in eukaryotic initiation factors that crosslinks specifically to the 5'-terminal cap in mRNA. Proc. Natl. Acad. Sci. USA 75, 4843-4847 (1978)

Sonenberg, N., Morgan, M.A., Testa, D., Colonno, R.J., Shatkin, A.J.: Interaction of a limited set of proteins with different mRNAs and protection of 5'-caps against pyrophosphatase digestion in initiation complexes. Nucleic Acdis Res. 7, 15-29 (1979a)

Sonenberg, N., Rupprecht, K.M., Hecht, S.M., Shatkin, A.J.: Eukaryotic mRNA cap binding protein: Purification by affinity chromatography on Sepharose-coupled m^7-GDP. Proc. Natl. Acad. Sci. USA 76, 4345-4349 (1979b)

Sonenberg, N., Trachsel, H., Hecht, S., Shatkin, A.J.: Differential stimulation of capped mRNA translation in vitro by cap binding protein. Nature (London) 285, 331-333 (1980)

Sonenberg, N., Guertin, D., Cleveland, D., Trachsel, H.: Probing the function of the eukaryotic 5' cap structure by using a monoclonal antibody directed against cap-binding proteins. Cell 27, 563-572 (1981a)

Sonenberg, N., Skup, D., Trachsel, H., Millward, S.: In vitro translation in reovirus- and poliovirus-infected cell extracts. Effects of anti-cap binding protein monoclonal antibody. J. Biol. Chem. 256, 4138-4141 (1981b)

Sonenberg, N., Guertin, D., Lee, K.A.W.: Capped mRNAs with reduced secondary structure can function in extracts from poliovirus-infected cells. Mol. Cell Biol. 2, 1633-1638 (1982)

Sonenberg, N., Edery, I., Darveau, A., Humbelin, M., Trachsel, H., Hershey, J.W.B., Lee, K.A.W.: Functional and structural characteristics of eukaryotic mRNA cap binding protein complex. In: Protein Synthesis (eds. A.K. Abraham, T.S. Eikhorn, J.F. Pryme). Clifton, N.Y.: The Human Press 1983

Steeg, H. van, Grinsven, M. van, Mansfeld, F. van, Voorma, H.O., Benne, R.: Initiation of protein synthesis in neuroblastoma cells infected by Semliki Forest virus. A decreased requirement of late viral mRNA for eIF-4B and cap binding proteins. FEBS Lett. 129, 62-66 (1981a)

Steeg, H. van, Thomas, A., Verbeek, S., Kasperaitis, M., Voorma, H.O., Benne, R.: Shutoff of neuroblastoma cell protein synthesis by Semliki Forest virus: Loss of ability of crude initiation factors to recognize early Semliki Forest virus and host mRNA's. J. Virol. 38, 728-736 (1981b)

Suzuki, H.: Effect of 7-methylguanosine-5'-phosphate on rabbit globin synthesis. FEBS Lett. 72, 309-313 (1976)

Suzuki, H.: Effect of $m^7G^{5'}ppp^{5'}Nm$ on the rabbit globin synthesis. FEBS Lett. 79, 11-14 (1977)

Tahara, S.M., Morgan, M.A., Shatkin, A.J.: Two forms of purified m^7G-cap binding protein with different effects on capped mRNA translation in extracts of uninfected and poliovirus-infected HeLa cells. J. Biol. Chem. 256, 7691-7694 (1981)

Tahara, S.M., Morgan, M.A., Grifo, J.A., Merrick, W.C., Shatkin, A.J.: Interactions of cap binding proteins with eukaryotic mRNAs. In: Interaction of Translational and Transcriptional Controls in the Regulation of Gene Expression (eds. M. Grunberg-Manago, B. Safer), Vol. XXIV, pp. 359-372. Amsterdam-New York: Elsevier Science Publishing Company 1982

Tahara, S.M., Morgan, M.A., Shatkin, A.J.: Binding of inosine-substituted mRNA to reticulocyte ribosomes and eukaryotic initiation factors 4A and 4B requires ATP. J. Biol. Chem. 258, 11350-11353 (1983)

Thomas, J.R., Wagner, R.R.: Inhibition of translation in lysates of mouse L cells infected with vesicular stomatitis virus: Presence of a defective ribosome-associated factor. Biochemistry 22, 1540-1464 (1983)

Trachsel, H., Erni, B., Schreier, M.H., Staehelin, T.: Initiation of mammalian protein synthesis. II. The assembly of the initiation complex with purified initiation factors. J. Mol. Biol. 116, 755-767 (1977)

Trachsel, H., Sonenberg, N., Shatkin, A.J., Rose, J.K., Leong, K., Bergmann, J.E., Gordon, J., Baltimore, D.: Purification of a factor that restores translation of vesicular stomatitis virus mRNA in extracts from poliovirus infected HeLa cells. Proc. Natl. Acad. Sci. USA 77, 770-774 (1980)

Van Steeg, H., Kasperaitis, M., Voorma, H.O., Benne, R.: Infection of neuroblastoma cells by Semliki Forest virus. Eur. J. Biochem. 138, 473-478 (1984)

Walden, W.E., Thach, R.E.: The role of mRNA competition in regulating translation in normal fibroblasts. In: Interaction of Translational and Transcriptional Controls in the Regulation of Gene Expression (eds. M. Grunberg-Manago, B. Safer), Vol. XXIV, pp. 399-416. Amsterdam-New York: Elsevier Science Publishing Company 1982

Webb, N.R., Chari, R.V.J., Depillis, G., Kozarich, J.W., Rhoads, R.E.: Purification of the messenger RNA cap-binding protein using a new affinity medium. Biochemistry 23, 117-181 (1984)

Weber, L.A., Feman, E.R., Hickey, E.D., Williams, M.C., Baglioni, C.: Inhibition of HeLa cell messenger RNA translation by 7-methylguanosine 5'-monophosphate. J. Biol. Chem. 251, 5657-5662 (1976)

Weber, L.A., Hickey, E.D., Nuss, D.L., Baglioni, C.: 5'-Terminal 7-methylguanosine and mRNA functions: Influence of potassium concentration on translation in vitro. Proc. Natl. Acad. Sci. USA 74, 3254-3258 (1977)

Weber, L.A., Hickey, E.D., Baglioni, C.: Influence of potassium salt concentrations and temperature on inhibition of mRNA translation by 7-methylguanosine 5'-monophosphate. J. Biol. Chem. 253, 178-183 (1978)

Wei, C.-M., Moss, B.: Methylated nucleosides block 5'-terminus of vaccinia virus messenger RNA. Proc. Natl. Acad. Sci. USA 72, 318-322 (1975)

Willems, M., Wieringa, B., Mulder, J., Ab, G., Gruber, M.: Translation of vitellogenin mRNA in the presence of 7-methylguanosine 5'-triphosphate. Cap analogs compete with mRNAs on the basis of affinity for initiation-complex formation. Eur. J. Biochem. 93, 469-479 (1979)

Wu, J.M., Cheung, C.P., Suhadolnik, R.J.: Differential inhibition with partially purified and endogenous rabbit reticulocyte globin mRNA by 7-methylguanosine 5'-monophosphate. Biochem. Biophys. Res. Commun. 78, 1079-1086 (1977)

Yamaguchi, K., Nakagawa, I., Sekine, M., Hata, T., Shimotohno, K., Hiruta, M., Miura, K.-I.: Chemical synthesis of the 5'-terminal part bearing cap structure of messenger RNA of cytoplasmic polyhedrosis virus (CPV): $m^7G^{5'}$pppAmpG and $m^7G^{5'}$pppAmpCpU. Nucleic Acids Res. 12, 2939-2954 (1984)

Zan-Kowalczewska, M., Bretner, M., Sierakowska, H., Szczesna, E., Filipowicz, W., Shatkin, A.J.: Removal of 5'-terminal m^7G from eukaryotic mRNAs by potato nucleotide pyrophosphatase and its effect on translation. Nucleic Acids Res. 4, 3065-3081 (1977)

Zarbl, H., Skup, D., Millward, S.: Reovirus progeny subviral particles synthesize uncapped mRNA. J. Virol. 34, 497-505 (1980)
Zumbé, A., Stahli, C., Trachsel, H.: Association of a M_r 50,000 cap-binding protein with the cytoskeleton in baby hamster kidney cells. Proc. Natl. Acad. Sci. USA 79, 2927-2931 (1982)

Cell-Cell Recognition in Lower Multicellular Eukaryotes. Models Supporting the Modulation Theory of Adhesion

W. E. G. Müller

A. Introduction

Little more than 100 years ago, Claus (1876) wrote *"Ebenso wenig wie die Zellmembran ist der Zellkern ein nothwendiger Charakter der Zelle. Es ist vielmehr das lebendige Protoplasma mit seiner nicht näher bekannten molekularen Anordnung das ausschließlich bestimmende Criterium der Zelle"*. Owing to the intense work of histochemists, cytochemists, and molecular biologists this view has been changed. Today it is well established that in eukaryotes the nucleus is filled with the genetic information for the synthesis of all proteins which function either as structural, functional, or controlling elements both in nucleus, cytoplasm, and in the cell membrane. On the other side, the cell membrane of unicellular and multicellular organisms is that structure which contains the receptors for extracellular signals. These "messengers" are predominantly involved in the control of the cell and tissue-specific readout of the genetic information. Moreover, cell membranes have acquired an additional, essential function during the phylogenetic evolution from the unicellular to the multicellular organisms; they carry the cell, tissue, and species-specific recognition system(s) and they are provided with mechanisms to recognize and destroy foreign cell entities. In animals, the cells are bound together by adhesion molecules which constitute a relatively loose meshwork of large extracellular organic molecules. In addition, cells of most multicellular eukaryotes are held together by intercellular junctions such as gap and tight junctions and desmosomes. Hence, such cell-cell interactions govern the positioning of cells within embryos during morphogenetic movements and establish a controlled cell-cell, cell-tissue, and tissue-organism interaction. These interactions are not static but dynamic and therefore require at least the following two mechanisms; firstly, the formation of stable linkages between cell surface molecules of adjacent cells and secondly, destruction of cell-cell binding by molecules and/or enzymic processes. The latter process is also known as "sorting out".

Besides the two separate kinds of cellular interaction (via cell-cell adhesion molecules and via intercellular junctions) a third mechanism has evolved to stabilize the position of a cell in a functional organism, the cell-substrate adhesion system. This involves the interaction of the cells with the extracellular matrix such as fibronectin, collagen, or laminin (Hay 1981).

B. Cell-Cell Adhesion in Vertebrates

It is conceivable that the elucidation of the mechanism(s) of cell-cell adhesion in vertebrates, with the aim to understand their role in histogenesis especially in the development of the nervous system, had at the beginning a higher priority than comparable studies with lower eukaryotes. However, due to the higher biological and biochemical complexity of the cell-cell recognition systems in higher eukaryotes and especially in vertebrates, the basic concept for the understanding of cell-cell recognition had been worked out in lower animals primarily using sponges. Today it is known that in vertebrates the following two distinct cell-cell adhesion systems exist.

I. The Ca^{2+}-independent Adhesion System

Based on the pioneering works of Moscona (Hausman et al. 1976), Lilien (Balsamo and Lilien 1974) and Glaser (Merrell et al. 1975), Edelman (1976, 1983) succeeded in the identification and purification of those molecules, which are involved in the Ca^{2+}-independent system. They were termed cell adhesion molecules (CAM). So far, they have been isolated from retina and brain (N-CAM), from liver (L-CAM) and from astrocyte-like glial cells with neurons (GN-CAM). The N-CAM is a large sialoglycoprotein with a M_r of 120,000. This molecule (single polypeptide chain) is present on cell surfaces of neurons and striated muscle precursors as an integral membrane protein. The N-CAM system has been associated with the formation of neurite bundles, the sorting out of neurites and somata in cell aggregates, and the organization of cell layers in retinal tissue (Brackenbury et al. 1981; Edelman 1983). Isolated N-CAM shows self-aggregation and microheterogeneity. The latter characteristics are attributed to its carbohydrate portion, which is composed predominantly of sialic acid. Immunological studies revealed, that N-CAM binding is second-order homophilic, which means that two identical CAM's form the linkage. Ca^{2+} is not required for this process. The CAM's have been shown to be functionally cell and tissue-specific but not species-specific (Chuong et al. 1982). Sialic acid itself is not directly involved in the adhesion process but, modulates its strength. In this context it is important to note that embryonic N-CAM has a sialic acid content of up to 26%, while adult molecules contain only 9% of this sugar. During development, in vivo N-CAM undergoes a conversion from sialic acid-rich to sialic acid-low. Concomitantly with this change, an alteration of the strength of the CAM-CAM interaction from low to high is assumed. This means, that sialic acid might modulate the free energy of binding between CAM molecules. During the transition of N-CAM from the sialic acid-rich to the sialic acid-low state the free energy strongly increases. The CAM molecules on the cell surface are mobile in the plane of the membrane and are characterized by a diffusion constant similar to that of the majority of the other surface receptors (Gall and Edelman 1981). These data support the modulation theory of adhesion (Edelman 1976), which had been substantiated first in sponges (Weinbaum and Burger 1973; Müller et al. 1976) and disfavour the chemoaffinity hypothesis (Sperry 1963), as well as the assumption that cell adhesion takes place via weak forces (Curtis 1967). The modulation theory requires only a few kinds of adhesion molecules in order to explain specific cell-cell interaction. In contrast, the chemoaffinity theory requires a multiplicity of different adhesion molecules which are pairwise complementary.

II. The Ca^{2+}-dependent, Trypsin-resistant Adhesion System

By application of a different tissue-dissociation technique (treatment with trypsin in the presence of Ca^{2+} by which CAM's are effectively destroyed) trypsin-resistant cell adhesion molecules (CAT's) have been detected and identified (Urushihara et al. 1979; Brackenbury et al. 1981). The following two distinct CAT systems can be distinguished, the Ca^{2+}-dependent mechanism in neural cells and the different Ca^{2+}-dependent mechanism in liver cells. These CAT systems and the CAM systems are antigenic distinct cell adhesion mechanisms (Brackenbury et al. 1981). The CAT's diplay little or no organ specificity. Referring to the neural system, inhibition of the CAT system causes an increase in the thickness of neurite fasciles, an effect not observed with antibodies against N-CAM. This result may indicate that the CAM and CAT systems play different roles during development (Brackenbury et al. 1981). They seem to act simultaneously during initial binding events in neural retina cells (McClay and Marchase 1982) and perhaps successively during embryogenesis (Brackenbury et al. 1981). So far, CAT's have not yet been solubilized from cell membrane and purified.

Recently, another Ca^{2+}-dependent adhesion system, which is responsible for compaction of the morula and of embryonal carcinoma cells, has been detected (Hyafil et al. 1980). It differs from the CAT system by the fact, that no trypsin treatment is required for its revelation. The underlying molecule, termed uvomorulin, has been isolated and purified (Hyafil et al. 1981).

C. Cell Adhesion Molecules in Sponges

Since the pioneering work of Wilson (1907) marine sponges afford the classic model for studies of cell-cell interactions on a molecular level. Later, Galtsoff (1929), who demonstrated the existence of a soluble aggregation factor, and Moscona (1963) contributed basic ideas to the understanding of cellular recognition in sponges. The detailed work on the underlying molecular mechanisms was performed by Humphreys (1970), Burger (1977), and Müller (1982).

Owing to their 1000 million years independent history, sponges have maintained some original features of early metazoans. Their cells are characterized by an unusually high reversible differentiation potency and by a comparably high motility. Because firstly, the junctions between sponge cells show only little

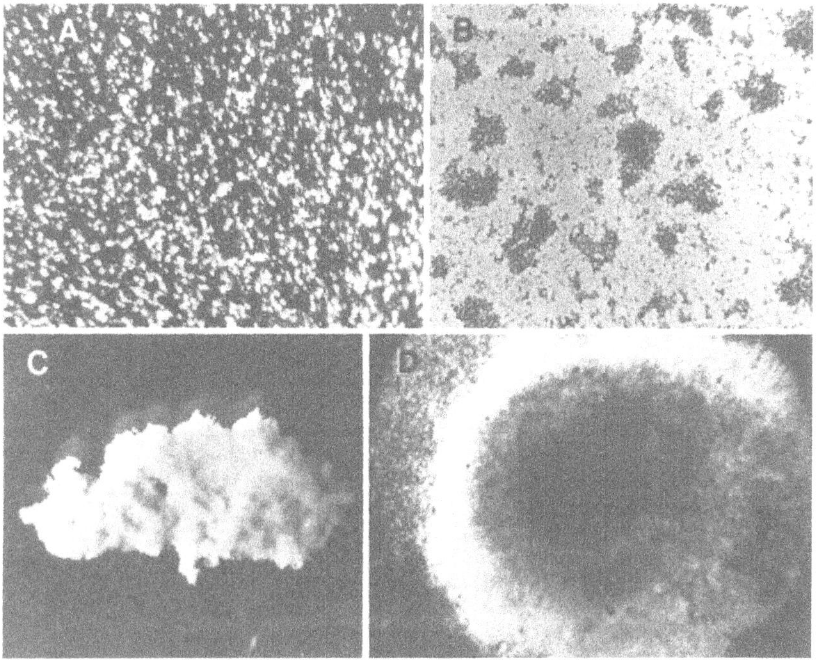

Fig. 1. Reaggregation of chemically dissociated cells from Geodia cydonium. A Single cells. x 200. B Primary aggregates. x 65. C Secondary aggregates. x 15. D Reconstituted organism. x 15. (From Müller 1982)

specialization, secondly, the distances between them are unusually high (between 300 and 1000 Å), and thirdly, the intercellular matrix into which they are embedded consists of a fluid phase (carbohydrate-rich compounds) and a solid phase (fibers, fibrils, and spicules), the sponge organism provides its cells with the structural prerequisites and the dynamic potentials by which they can arrange themselves in a highly flexible and dynamic way in tissue units. For the stabilization of such a multicellular system, highly developed cell-cell recognition mechanisms are present in sponges which enable their cells to contact inhibition of movement, contact promotion, and to have contact guidance. Two distinct recognition systems have been elucidated; both of them are Ca^{2+}-dependent. In the following, we discuss primarily our own work, performed with the *Tetractinomorphans* (siliceous sponge) *Geodia cydonium* (Jam.) and *Cliona celata* (Grant).

I. Ca^{2+}-dependent Primary Aggregation System

Cells of marine sponges, belonging to the classes Hexactinellida (unpublished), Calcarea (Müller 1982) and Demospongia (Müller 1982) are provided with the primary Ca^{2+}-dependent aggregation system. Detailed work with *Geodia cydonium* (Müller and Zahn 1973; Müller et al. 1974, 1985) and *Cliona celata* (α-stage) (Müller et

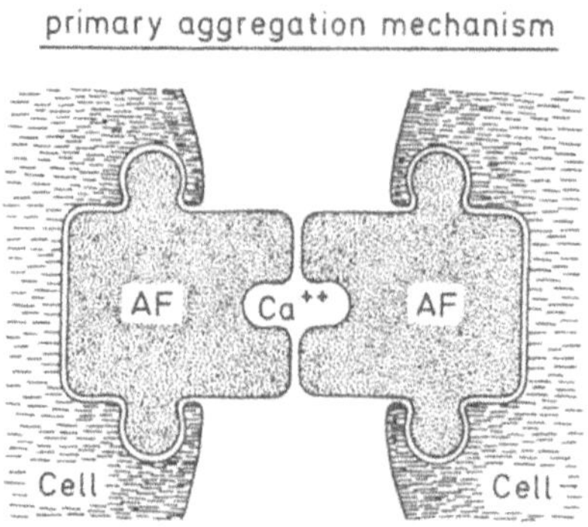

'ig. 2. Diagrammatic representation of :he two molecular mechanisms proposed :o explain primary and secondary aggre- ation mechanism in the sponge *Geodia* *ydonium*. *AF* aggregation factor; *AR* ag- regation receptor; *Ca*$^{++}$ calcium ion; ' core structure

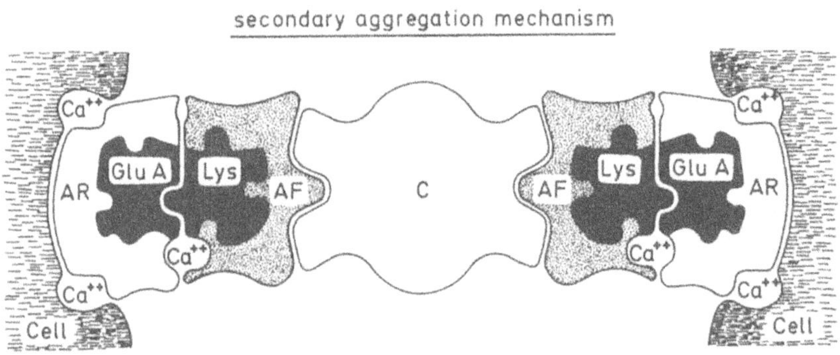

al. 1982c) revealed, that at least 18 mM of Ca^{2+} are required for formation of 80 - 500 µm large aggregates (Fig. 1A and B).

By an immunochemical approach it could be demonstrated that primary aggregation in *Geodia cydonium* is mediated by a specific ligand(s) which was termed primary aggregation factor (pAF) (Müller et al. 1985). The pAF is firmly associated with the cell surface (Fig. 2). Solubilizers as Triton X-100, Nonidet NP-40, Tween 80, n-octyl-βD-glucopyranoside, or lithium,3,5-diiodosalicylate failed to remove the pAF from the membrane. Only after incubation of the cells with urea the pAF was recovered in a soluble form. After purification almost to homogeneity, the factor turned out to be composed of three subunits with the following mol. wts.: 16,000, 15,000 and 13,500. The pAF is a glycoprotein (48% protein, 34% neutral carbohydrates, 13% hexuronic acid, 4% hexosamines) which is heat-labile and rapidly destroyed by proteases. This purified pAF is still provided with the property of binding to surfaces of both homologous and heterologous urea-treated sponge cells. As known so far, pAF-mediated binding is second-order homophilic (pAF-pAF). No data on the molecular events proceeding or during pAF-pAF interaction are available.

An important observation, obtained first with the *Cliona*-system (Müller et al. 1982c) and later with *Geodia cydonium* (Müller et al. 1985), revealed that the primary aggregation process occurs, at least during the initial phase, non-species specific. For

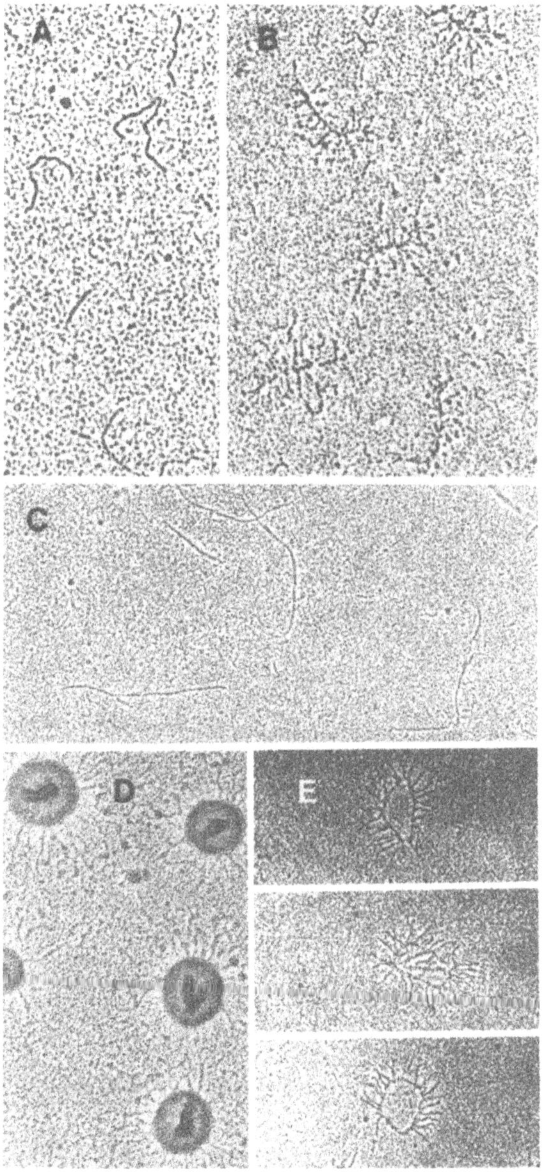

Fig. 3. Electron micrographs of aggregation factors. A *Cliona celata:* Rod-like structure from the α-stage. x 60,000. B *Cliona celata:* Rod-like structure with side chains from the γ-stage. x 50,000. C *Clathrina coriacea* (Mont.; Calcarea: Clathrinida). x 28,000. D *Geodia cydonium:* Native aggregation factor. x 70,000. E *Geodia cydonium:* Core structure. x 55,000. (A and B from Müller et al. 1982c; D and E from Müller 1982)

example, for the species combination *Cliona-Geodia* a nearly complete intermingling of cells in the aggregate takes place. This fact raises the question for the mechanism by which sponges, which are provided with the primary aggregation system (e.g., all Calcarians), maintain their species-identity. As outlined later, the secondary aggregation system (present in most Demosponges) is species-specific. Therefore, it was very unexpected to find that the Demosponge *Cliona celata* can develop and grow both in the absence and in the presence of the secondary aggregation factor, depending on the growth state of this species (Müller et al.

1982c). This sponge exists in three growth stages; α-stage: boring form, β-stage: transition form (overgrowing the surface of the substratum) and γ-stage: crust form (living entirely on the substratum). In the α-stage, the sponge is provided only with the primary aggregation mechanism, while cells from the crust form (γ-stage) reaggregate on the basis of the two Ca^{2+}-dependent aggregation principles. The secondary aggregation factor from γ-stage material was found to be associated with rod-shaped structures of 0.3 μm of length which have approximately 20 side chains (0.06 μm); Fig. 3B. The core of this structure, plain rods (Fig. 3A) without side chains, is present in α-stage material but obviously not associated with the secondary aggregation factor. Therefore, we can ask the question whether these large structures become, after they attain the appropriate size, a vehicle for the secondary aggregation factor, located elsewhere at a hitherto unknown position. If this is true, we might have overlooked the "secondary aggregation factor" in α-stage *Cliona* and also in Calcarians. Some indications seem to support this assumption; e.g., the secondary aggregation factor from *Geodia* can be solubilized from large carrier structures (Müller et al. 1974) and those sponge species hitherto considered as organisms lacking a secondary aggregation mechanism (*Clathrina coriacea*, Fig. 3C) show structures similar to the characteristically large secondary aggregation factor complex (Müller et al. 1982a).

II. Ca^{2+}-dependent Secondary Aggregation System

Most Demosponges (see: Müller 1982) and Hexactinellidians (*Aphrocallistes vastus*; unpublished) are provided, in addition to the primary cellular recognition system, with a secondary adhesion apparatus. Basically, it functions in the same way in the different models used by Humphreys (Henkart et al. 1973), Burger (Weinbaum and Burger 1973) and us (Müller and Zahn 1973; Müller et al. 1976). We worked primarily with *Geodia cydonium*, a siliceous sponge weighing up to 50 kg.

The secondary cell adhesion system is characterized by the presence of the secondary aggregation factor molecule (sAF), which is readily obtained in a soluble form after incubation of *Geodia* tissue in isotonic solution, containing 20 mM EDTA. Re-addition of the sAF to single cells results in the formation of large aggregates (5 mm and larger) (Fig. 1C). After incubation for several days a rearrangement of the cells in the aggregates to a functioning sponge (Fig. 1D) occurs (Müller and Zahn 1973). The sAF is a complex macromolecule with a sedimentation coefficient of 90. In its "native" form the sAF appear as spheres (diameter: 1000 Å) with a concave cup structure (Fig. 3D). The center is electron dense after platinum shadowing. After Nonidet-treatment of the *Geodia* particle, a core structure is obtained which appears as a "sunburst" with a circular center (diameter: 1100 Å) and 25 radiating arms (Fig. 3E). The core structure of the sAF from some other sponges is rod-like and carries side chains (Müller et al. 1982a). The *Geodia* sAF is intracellularly assembled and becomes firmly associated with the cell membrane during the release into the extracellular space (Müller et al. 1982b). There, the sAF is embedded into the dense intercellular matrix and establishes cell-cell contact. The *Geodia* sAF is provided with the following three functional subunits (Müller et al. 1978d); the glucuronosyltransferase (which is involved in the "sorting out" process), the galactosyltransferase (of unknown function) and that subunit which carries the binding site to the cell membrane. The latter subunit (the low-molecular AF) is not covalently linked with the core structure (Müller et al. 1974) (Fig. 2).

The low-molecular AF-subunit, which consists of 91% protein and 6 % carbohydrate binds to a specific molecule on the cell surface, the aggregation receptor (AR) (Weinbaum and Burger 1973; Müller et al. 1976) (Fig. 2). The chemical groups which are involved in the recognition of the AR by the AF are L-lysine (in the AF molecule) (Müller et al. 1974) and glucuronic acid (in the AR unit) (Vaith et al. 1979a). Ca^{2+} plays an important role also during the secondary aggregation process; its binding sites are not yet known. While in the *Microciona* system Ca^{2+} is apparently involved in sAF-sAF polymerization (Burkart and Burger 1981), this cation functions in the *Geodia* system during the binding processes of AR to the cell surface (Müller et al. 1976) and sAF to AR (Müller et al. 1974; Fig. 2). Common to all sponge models investigated (Burkart and Burger 1981, Müller 1982) is the fact, that adhesion between cells, which bases on the secondary aggregation system, is third-order heterophilic (AR-sAF-AR); AR and sAF are different CAM's (in the nomenclature of Edelman 1983) and sAF is present as binding ligand.

Since early biochemical studies (Humphreys 1963; Moscona 1963), it was well established that species-specific reaggregation of sponge cells is due to the action of the species-specific sAF, provided that it is sufficiently pure (Müller et al. 1980), which means that the sAF is free of lectin. Two categories of specificity are distinguished: first, partial species-specificity realized in the system *Haliclona occulata* - *Microciona prolifera* (Turner and Burger 1973) and second, a total species-specificity as described for the combination *Geodia cydonium* - *Suberites domuncula* (Müller et al. 1978a, 1978c). The results obtained from experiments with the latter system confirm in a clear manner the modulation theory of adhesion: From these two sponge species both the sAF and the AR were isolated and purified. In the absence of any factor, cells from these species form primary aggregates only with a diameter less than 100 μm. After addition of sAF to homologous cells, they respond and form secondary aggregates. The reciprocal experiments revealed no response of *Geodia* cells to *Suberites*-sAF and no secondary aggregation of *Suberites* cells to *Geodia*-sAF. This means that these two sAF's recognize even in the initial phase heterologous cells and refuse to aggregate with them. However after preincubation of *Geodia* cells with *Suberites*-AR and subsequent incubation with *Suberites*-sAF large aggregates are formed. This means that *Geodia* cells which have been charged with *Suberites*-AR behave like *Suberites* cells in the presence of *Suberites*-sAF. From this study it seems to be obvious that species-specific aggregation is due to a qualitative difference in the reactivity of the sAF with the AR. The biochemical data demonstrate a perfect interaction between sAF and AR and have recently been supported by immunological studies as well (Conrad et al. 1981). Consequently,

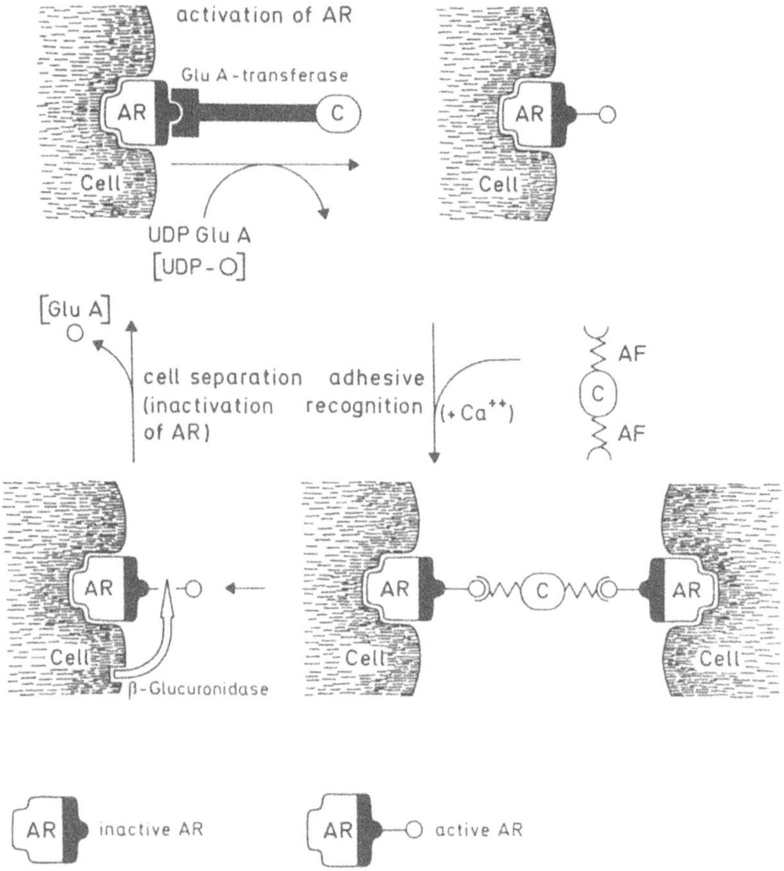

Fig. 4. Proposed molecular mechanism for cell aggregation and cell separation via activation or inactivation of the aggregation receptor *(AR)*. *C* core structure; *AF* secondary aggregation factor; *GluA* and *o* glucuronic acid

we have a pronounced specificity of secondary adhesion within an organism which is based on only two kinds of CAM's, the sAF and the AR. The "activity" of these molecules is modulated by the following enzymic and nonenzymic processes.

1. Modulation by Enzymic Interactions

The *Geodia* system served as an appropriate model for the biochemical investigation of the reversible transition from aggregation-susceptible- to aggregation-deficient cells (the basis for a controlled cell movement and positioning). The ability of the cells to form aggregates is controlled by the binding capacity of the AR (Müller et al. 1979c). As mentioned above, the AR carries a glucuronic acid moiety in a terminal position. This acidic sugar can, under defined conditions, be removed from the AR by the cell surface-bound β-glucuronidase, resulting in an inactivation of its binding function. The biological consequence of this enzymic process is cell separation (Fig. 4). The deglucuronylation is a reversible event. Catalyzed by the sAF-associated glucuronosyltransferase the glucuronic acid unit can be reattached to the AR by which this molecule is reactivated. After this process the cells have regained their aggregation potency. This proposed mechanism of an enzymically controlled cell adhesion, as supported by experimental data (Müller et al. 1979c) has some striking similarities with the Roseman theory proposed 1970 (Roseman 1970). At present, we have no experimental evidence on the mechanisms by which glucuronylation and deglucuronylation are controlled. One possibility could be an alteration in the cell surface pH micro-milieu. The glucuronosyltransferases, associated with the sAF, have pH optima of 8.2 (Müller et al. 1978d), while the cell membrane-bound glycosidases are active around pH 5.0 (Müller et al. 1979c). No data are available which give direct experimental evidence for a pH or ion change of the sponge cell membrane during the process of cell separation. However from previous studies (Müller et al. 1978d), it is known that the sAF from *Geodia* alters the cell surface charge after its binding to homologous cells. Because alterations of the cell surface charge can also be caused by a pH change at the cell surface (Curtis 1967), we postulate that pH fluctuations govern the activity of both the glycosidases and glycosyltransferases.

A very similar modulation mechanism for cell adhesion molecules has recently been described for the N-CAM system from higher eukaryotes (Rothbard et al. 1982). The authors propose that the fluctuation of sialic acid content of N-CAM during development might be enzymically controlled (Edelman 1983) by a balanced glycosylation-deglycosylation mechanism.

2. Modulation by Nonenzymic Interactions

A second aggregation-regulating system has been detected on the surface of sponge cells (Müller et al. 1979a). The key molecule in this system is the anti-aggregation receptor, which can bind with its terminal galactose to the sAF-AR complex (Müller et al. 1979b). As consequence of the binding, this complex is destroyed and the cells separate from each other (Fig. 5). The function of the anti-aggregation receptor is controlled by a β-galactose-specific lectin which has the property to reversibly mask the terminal galactose of the anti-aggregation receptor. This lectin, originally described by Vaith et al. (1979b), has some unique properties (Müller et al. 1983b). It is a trimer (subunits: M_r 13,800, 13,000 and 12,200) in which two subunits are linked by one disulfide bond. All three subunits carry carbohydrate side chains with terminal D-galactose residues in β1 → 6 and/or β1 → 4 glycosidic linkages. This means, that the *Geodia* lectin has, in addition to carbohydrate recognition sites for D-galactose (two sites per molecule), at least one receptor site for a second lectin molecule. This "self-recognizing" *Geodia* lectin was experimentally shown to bind to the anti-aggregation receptor (Müller et al. 1979a). Moreover, it could be demonstrated by affinity chromatography (Müller 1982) that in the absence of Ca^{2+} the sAF binds to the anti-aggregation receptor and not to the AR. From these experiments, the scheme shown in Figure 5 has been proposed.

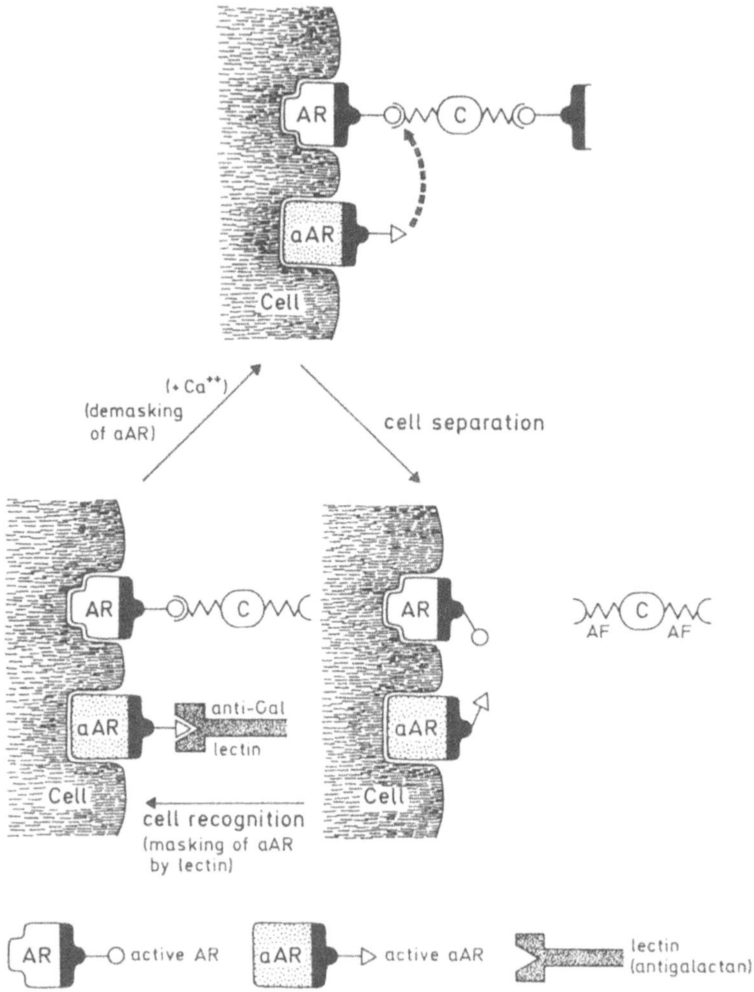

Fig. 5. A tentative model for the biological role of the *Geodia cydonium* lectin during the reaggregation of homologous cells. *AR* aggregation receptor; *C* core structure; *AF* secondary aggregation factor; *aAR* anti-aggregation receptor; ▷ galactose

The conclusion that the lectins are involved in the control of cell adhesion is not restricted to sponges. A definite role of lectins in this process has been established in slime molds (Barondes and Rosen 1976); a comparable function in vertebrates is very likely (Kitamura 1980). The static model, presented in Figure 5, does not explain, e.g., the origin of the lectin and the mechanism by which this molecule is guided to the anti-aggregation receptor. Regarding the dynamic situation, it could be shown that the *Geodia* lectin is synthesized in mucoid or spherulous cells (Müller et al. 1981b; Bretting et al. 1982). From immunological and biological studies it was concluded (Müller et al. 1981b) that the lectin synthesized in the mucoid cells is secreted and subsequently transferred to the cell surface of other cell types. Quantitative studies revealed that a total of 9.1×10^6 lectin molecules are present per mucoid cell. As one consequence of the binding of the lectin to cell surface, a conversion of aggregation-deficient choanocytes or achaeocytes to aggregation-susceptible ones was observed. In this context it is again interesting, that in slime molds (Springer et al. 1980) a very similar lectin localization and dissemination mechanism exists.

We assume that the two described mechanisms, modulating secondary aggregation process, are involved in cell movement and "sorting out" (Müller 1982). These events are essential for a reconstitution of randomly aggregated cell clumps to a functional organism. During this reconstitution process the cells must be provided with the following abilities: contact inhibition of movement when two locomoting cells make contact, chemotactic orientation, and oriented movement on a substrate. It is not known, whether the movement of sponge cells is guided by a gradient of morphogenetic substances, or by the extracellular skeleton of proteinaceous and glycoproteinaceous nature or by both mechanisms.

3. Modulation by Histoincompatibility Reactions

Besides the tissue- and cell-specific recognition systems localized on the cell surface, the sponges have – as all other multicellular organisms – developed a host defense mechanism against foreign invaders. These mechanisms enable them to distinguish "self" from "nonself". To elucidate the molecular basis for individuality in sponges, graft rejection studies between allogeneic and xenogeneic individuals were successfully applied. Using the following species *Amphilectus fucorum* × *Hymeniacidon perleve* (Evans and Curtis 1978) and *Geodia cydonium* × *Geodia rovinjensis* (Müller et al. 1981a, 1983c) the process of xenograft rejection could be subdivided into the following three events (a) fusion of the contact areas, (b) necrotic alteration of the graft and (c) rejection of the xenogeneic graft (Fig. 6). Immediately after fusion, a cytotoxic material (termed inhibitory aggregation factor) is synthesized by the graft (Evans and Curtis 1978; Müller et al. 1981a) which causes its own death. This means, that the graft receives a signal and destroys itself by a lytic action. This "induced suicide" mechanism, first observed in gorgonians (Theodor 1970) seems to be a general histoincompatibility reaction in the earliest metazoans. The purified inhibitory aggregation factor from *Geodia cydonium* is of low mol. wt. (M_r = 27,000) and glycoproteinaceous in character (Müller et al. 1981a). Kinetic studies revealed that the inhibitory aggregation factor reduces the process of secondary aggregation by competing with the AR for the binding sites at the sAF. In other words, the inhibitory aggregation factor abolishes and thus modulates the activity of the sAF by a reversible masking of its binding sites for AT-molecules on the cell surfaces. These observations, revealing that the synthesis of the inhibitory aggregation factor is induced in those cells which have lost their physiological environment and their contact to isogeneic cells, indicate that sponges are provided with a well-developed tissue immunopotential. This

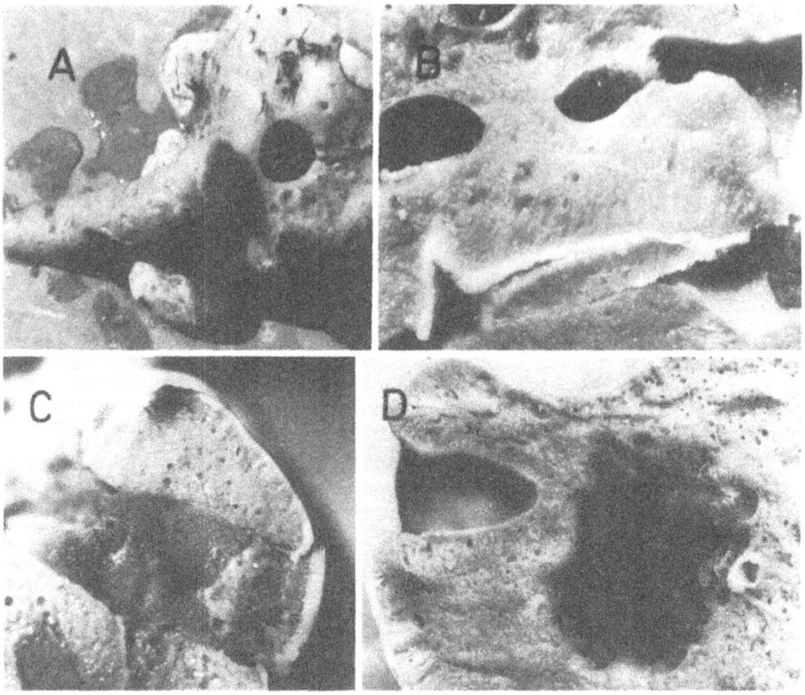

Fig. 6. Graft experiments with *Geodia cydonium* and *Geodia rovinjensis*. A Insertion of a *G. rovinjensis* graft into a hole, punched into *G. cydonium*. x 0.5. B Fusion of an allograft. x. 1. C Rejection of a xenograft from *G. rovinjensis* in *G. cydonium*. x 1. D Cytotoxic reaction in *G. cydonium* caused by the inhibitory aggregation factor, isolated from xenografts of *G. rovinjensis*. x 2. (From Müller 1982)

system is simple and efficient; it involves a further molecule, the inhibitory aggregation factor, which interferes with the third-order heterophilic secondary aggregation system.

As outlined by Turner (1978), grafting experiments are not suitable to elucidate allogeneic immunopotentials in sponges. Using a slightly different approach by placing different individuals into close contact, Hildemann et al. (1979a,b) were able to show that sponges are already provided with an allogeneic immunocompetence. In contrast to the processes during xenograft rejection, the cytotoxic reactivity occurring during allogeneic reaction does not involve a readily diffusible cytotoxic molecule. It would be highly interesting to know whether allogeneic reactions proceed by modulation of the second-order heterophilic primary aggregation system. It should be noted, that the highly discriminating xenogeneic and allogeneic immunocompetence comprises an inducible specific memory component (Hildemann et al. 1979b).

D. Cell Recognition Molecules in Corals

Like all other multicellular organisms, the coelenterates have developed a recognition system which is localized on the cell periphery, in order to establish and to stabilize firstly, the functional arrangement of cells in the organism and secondly, the distinction between syngeneic, allogeneic, and xenogeneic tissue. The latter aspect had been studied in hydrozoans (Ivker 1972; Müller et al. 1983a), alcyonarians (Theodor 1970), and zoantharians (Hildemann et al. 1977; Lubbock et al. 1980). We have approached the elucidation of a cell-cell adhesion mechanism in a coelenterate on molecular level and have chosen the gorgonians *Eunicella cavolinii* and *Echinomuricea coccinea* (Müller et al. 1983b, submitted).

The results obtained (Müller et al. 1983d) are in agreement with the supposition that specific aggregation molecules mediate cell-cell contact also in the *Eunicella* system. The aggregation molecules are firmly associated with the cell surface and are not removed from it by 0.75 M urea or EDTA. They are resistant to trypsin. Moreover, interaction between aggregation molecules from opposite cells occurs in the absence of Ca^{2+}. Hence, this mechanism resembles the Ca^{2+}-independent system in Chinese hamster V79 cells (Urushihara et al. 1979) and in neural cells (Brackenbury et al. 1981). Until now we failed to solubilize the aggregation molecule from this coral. The 0.4 - 2.3 µm long pearl cord-like structures, which are present in the intercellular matrix (Fig. 7B) and resemble the sAF of the sponge *Ircinia muscarum* (Müller et al. 1982a) are not involved in cell-cell recognition process.

The cell-free supernatant from *Eunicella* contains large amounts (7.3% of the total protein) of a galactose-specific lectin. The evidence for its existence was first obtained (Müller et al. 1983d) both from double-diffusion experiments and from hemagglutination inhibition studies. Its agglutinating activity could not be blocked by mono- or disaccharides, indicating that the *Eunicella* lectin recognizes besides galactose its configuration at the anomeric carbon atom as well. High-affinity reactions were determined between this antigalactan and a series of higher mol. wt. galactans; the highest values were calculated for the *Littorina littorea* snail galactan with an inhibition titer of 2^{13} (hemagglutination inhibition assay). Recently we succeeded in the purification of the *Eunicella* lectin. The crude extract was fractionated by ion exchange chromatography on DEAE cellulose (Fig. 8). The lectin was eluted with 1 M NaCl and determined to be electrophoretically to be nearly homogeneous (Fig. 7A). Under denatured conditions the lectin showed an apparent mol. wt. of 23,400. Gel filtration experiments using Biogel P 300 revealed (unpublished) for the native lectin a mol. wt. of 95,000. This result is compatible with the assumption that the *Eunicella* lectin is a tetramer. It contains in all subunits, carbohydrates as deduced from the periodic acid/Schiff positive reaction of the band after gel chromatography (Fig. 7A, a and b).

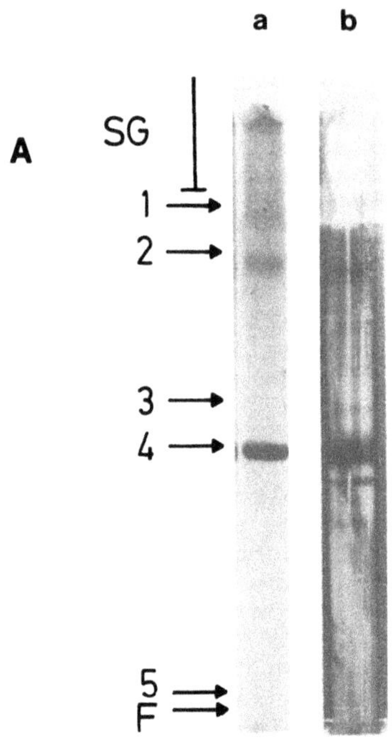

Fig. 7. A Gel electrophoresis of purified *Eunicella* lectin in 15% SDS polyacrylamide gels in the presence of 0.1% 2-mercaptoethanol; 10 µg of protein was applied per gel. *a* Coomassie blue staining; *b* periodic acid/Schiff staining. The *arrows* mark the positions of the following protein standards, which were run on parallel gels: *1* bovine serum albumin (M_r = 68,000); *2* ovalbumin (M_r = 43,000); *3* concanavalin A (promoter: M_r = 25,500); *4* trypsin inhibitor (M_r = 21,500); *5* cytochrome c (M_r = 12,500). *SG* stacking gel; *F* front. Further details were given earlier (Müller et al. 1983b). B Pearl cord-like structures, present in the intercellular matrix of *Eunicella cavolinii*. Electron micrograph; stained with uranyl acetate and lead citrate. x 24,000. (Photo courtesy of R.K. Zahn)

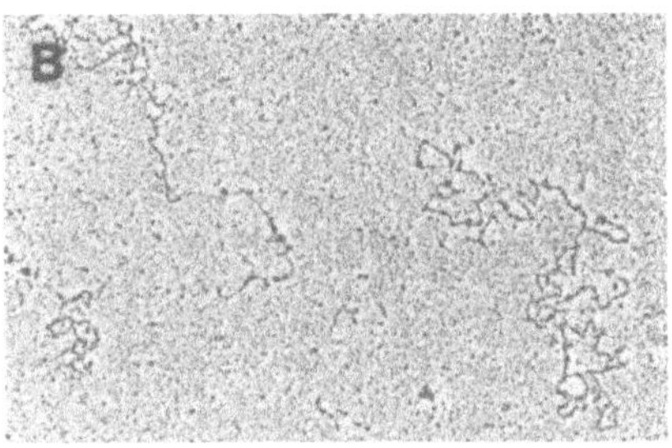

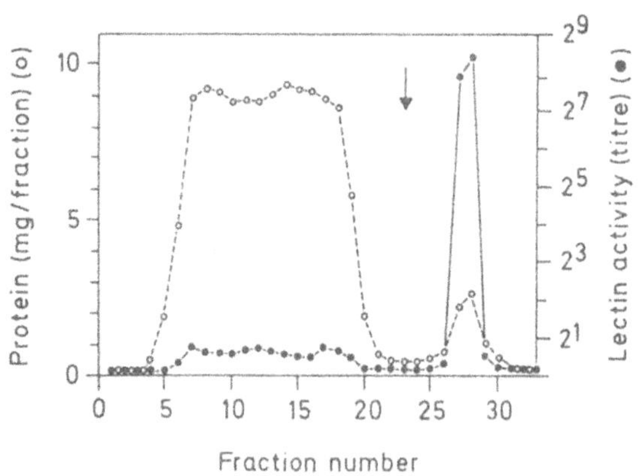

Fig. 8. Purification of the lectin from *Eunicella cavolinii* by ion exchange chromatography. A column (1.5×16cm) of DEAE-cellulose (DE 52), equilibrated with 100 mM Tris-HCl (pH 8.2; 300 mM NaCl), was loaded with 57 ml of crude lectin fraction (Müller et al., submitted) containing 131.1 mg of protein. Fractions of 6 ml were collected. After washing with the equilibration buffer, the lectin was eluted with 100 mM Tris-HCl (pH 8.2; 1 M NaCl); *arrow*. Aliquots of the fractions were assayed for lectin activity (●——●) and protein content (o---o). The lectin activity was determined in a haemagglutination assay. (Vaith et al. 1979b)

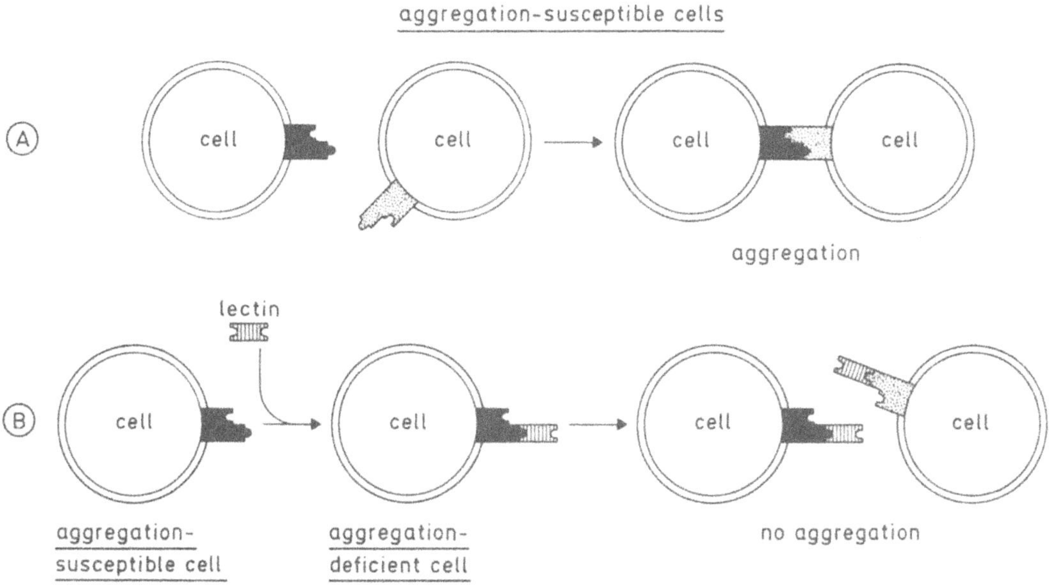

Fig. 9. A model showing the proposed cell-cell interaction in *Eunicella* via cell surface-bound aggregation molecules *(closed* and *dotted symbols)*. A Recognition. B Rejection due to the binding of the lectin (antigalactan; *hatched symbol*) to the aggregation molecule (galactan)

It was interesting to find (Müller et al. 1983d), that the galactose-specific *Eunicella* lectin blocks the aggregation potency of the homologous cells. This inhibition could be abolished by coincubation with an edible birdnest glycoprotein. Such a lectin function during cell-cell recognition was up to now unknown. Usually a lectin may support and/or allow adhesion of cells, e. g., in sponges (see above), slime molds (Drake and Rosen 1982), or sea urchin embryos (Tonegawa 1982). Hence, the lectin in the coral system does not act via a lock and-key mechanism, in which lectins and their carbohydrate-containing receptors mediate specific cell-cell recognition. Unlike that stated in the lectin hypothesis of cell adhesion (Barondes 1981), the *Eunicella* lectin converts aggregation-susceptible cells to aggregation-deficient cells and thus controls cell-cell interaction as a negative modulator (Fig. 9). It is assumed in the new model (Fig. 9), that in spite of its multivalency, one *Eunicella* lectin molecule can only bind to one aggregation molecule, resulting in a masking of its aggregative function. This would mean, that this lectin does not act as a bridging molecule, bringing two cell surfaces into direct contact. Such a lectin is termed an "incomplete" lectin, an expression, which has been adopted from the Rhesus blood group serology (Uhlenbruck 1971) for nonagglutinating, receptor-blocking and often also non-precipitating antibodies. Whether a lectin is "incomplete" or not, is in most cases due to the cryptic location of the receptor within the membrane architecture (Uhlenbruck 1971). Detailed biochemical work is now required to substantiate this proposed model. In this connection, special interest will be given to the question, whether other galactose-specific lectins of the antigalactan type can replace the *Eunicella* lectin in the aggregation model, or if they may cause instead of a blocking-induced "no aggregation" a galactose-lectin-induced "agglutination".

The data available with Gorgonians indicate that only one species of molecules is involved in cell-cell adhesion and therefore the system has to be classified to the group of second-order homophilic ones.

E. Conclusions

It is amazing but still plausible that in all animal systems hitherto studied there is only one basic principle by which cell-cell recognition occurs. From sponges and corals to higher eukaryotes there is only a small number of cell adhesion molecules which are involved in cell-cell adhesion. Specificities and histogenetic patterning are achieved by modulating processes acting on cell adhesion molecules. In sponges (secondary aggregation system) and in eukaryotes, the activity and function of cell adhesion molecules are assumed to be modulated by enzymic processes (glycosylation and deglycosylation). Strong evidence from experiments with corals and sponges is available which indicates that modulation can also occur nonenzymically by reversible interaction of cell adhesion molecules with lectins.

Acknowledgements. This work was supported by grants of the "Deutsche Forschungsgemeinschaft (Mu 348/8-1) and of the Bundesministerium für Forschung und Technologie (German-Yugoslavic cooperation program).

References

Balsamo, J., Lilien, J.: Functional identification of three components which mediate tissue-type specific adhesion. Nature (London) 251, 533-534 (1974)

Barondes, S.H.: Lectins: Their multiple endogenous cellular functions. Annu. Rev. Biochem. 50, 207-231 (1981)

Barondes, S.H., Rosen, S.D.: Cell surface carbohydrate-binding proteins: Role in cell recognition. In: Neuronal Recognition (ed. S.H. Barondes), pp. 331-356. New York: Plenum Press 1976

Brackenbury, R., Rutishauser, U., Edelman, G.M.: Distinct calcium-independent and calcium-dependent adhesion systems of chicken embryo cells. Proc. Natl. Acad. Sci. USA 78, 387-391 (1981)

Burger, M.M.: Mechanisms of cell-cell recognition: Some comparisons between lower organisms and vertebrates. In: Cell Interactions in Differentiation (eds. M. Karkinen-Jääskeläinen, L. Saxén, L. Weiss), pp. 357-376. London-New York: Academic Press 1977

Burkkart, W., Burger, M.M.: The contribution of the calcium-dependent interaction of aggregation factor molecules to recognition: A system providing additional specificity forces. J. Supramol. Struct. Cell. Biochem. 16, 179-192 (1981)

Bretting, H., Jacobs, G., Donadey, C., Vacelet, J.: The lectin-producing cells in the sponge tissue: In: Lectins-Biology, Biochemistry, Clinical Biochemistry (ed. T.C. Bog-Hansen), Vol.II, pp. 91-103. Berlin: de Gruyter 1982

Chuong, C.M., McClain, D.A., Streit, P., Edelman, G.M.: Neural cell adhesion molecules in rodent brains isolated by monoclonal antibodies with cross-species reactivity. Proc. Natl. Acad. Sci. USA 79, 4234-4238 (1982)

Claus, C.: Grundzüge der Zoologie. Marburg-Leipzig: Elwert 1876

Conrad, J., Zahn, R.K., Kurelec, B., Uhlenbruck, G., Müller, W.E.G.: Aggregation of sponge cells: Immunological characterization of the species-specific aggregation factor. J. Supramol. Struct. Cell. Biochem. 17, 1-9 (1981)

Curtis, A.S.G.: The Cell Surface: Its Molecular Role in Morphogenesis, p. 70. London-New York: Academic Press 1967

Drake, D.K., Rosen, S.D.: The pallidin-receptor system in the intercellular adhesion of the cellular slime mold, Polysphondylium pallidum. Cellular Recognition, pp. 777-787. New York: Liss 1982

Edelman, G.M.: Surface modulation in cell recognition and cell growth. Science 192, 218-226 (1976)

Edelman, G.M.: Cell adhesion molecules. Science 219, 450-458 (1983)

Evans, C.W., Curtis, A.S.G.: Graft rejection in sponges: Its relation to cell aggregation studies. In: Biologie des Spongiaires. Colloq. Int. C. N. R. S. 291, 211-215 (1978)

Gall, W.E., Edelman, G.M.: Lateral diffusion of surface molecules in animal cells and tissues. Science 213, 903-905 (1981)

Galtsoff, P.S.: Heteroagglutination of dissociated sponge cells. Biol. Bull. Mar. Biol. Lab. Woods Hole 57, 250-260 (1929)

Hausman, R.E., Knapp, L.W., Moscona, A.A.: Preparation of tissue-specific cell aggregating factors from embryonic neural tissues. J. Exp. Zool. 198, 417-422 (1976)

Hay, E.D.: Cell Biology of Extracellular Matrix. New York: Plenum Press 1981

Henkart, P., Humphreys, S., Humphreys, T.: Characterization of sponge aggregation factor. A unique proteoglycan complex. Biochemistry 12, 3045-3050 (1973)

Hildemann, W.H., Raison, R.L., Cheung, G., Hull, C.J., Akaka, L., Okamoto, J.: Immunological specificity and memory in a scleractinian coral. Nature (London) 270, 219-223 (1977)

Hildemann, W.H., Bigger, C.H., Johnston, I.S.: Histoincompatibility reactions and allogeneic polymorphism among invertebrates. Transplant. Proc. 11, 1136-1142 (1979a)

Hildemann, W.H., Johnston, I.S., Jokiel, P.L.: Immunocompetence in the lowest metazoan Phylum: Transplantation immunity in sponges. Science 204, 420-422 (1979b)

Humphreys, T.: Chemical dissolution and in vitro reconstitution of sponge cell adhesions. I. Isolation and functional demonstration of the components involved. Dev. Biol. 8, 27-47 (1963)

Humphreys, T.: Biochemical analysis of sponge cell aggregation. Symp. Zool. Soc. London 25, 325-334 (1970)

Hyafil, F., Morello, D., Babinet, C., Jacob, F.: A cell surface glycoprotein involved in the compaction of embryonal carcinoma cells and cleavage stage embryos. Cell 21, 927-934 (1980)

Hyafil, F., Babinet, C., Jacob, F.: Cell-cell interactions in early embryogenesis: A molecular approach to the role of calcium. Cell 26, 447-454 (1981)

Ivker, F.B.: A hierarchy of histo-incompatibility in Hydractinia echinata. Biol. Bull. 143, 162-174 (1972)

Kitamura, K.: The changes in lectin activity during the development of embryonic chick skin. J. Embryol. Exp. Morphol. 59, 59-69 (1980)

Lubbock, R.: Clone-specific cellular recognition in a sea anemone. Proc. Natl. Acad. Sci. USA 77, 6667-6669 (1980)

McClay, D.R., Marchase, R.B.: Calcium-dependent and calcium-independent adhesive mechanisms are present during initial binding events of neural retina cells. J. Cell. Biochem. 18, 469-478 (1982)

Merrell, R., Gottlieb, D.I., Glaser, L.: Embryonal cell surface recognition. Extraction of an active plasma membrane component. J. Biol. Chem. 250, 5655-5659 (1975)
Moscona, A.A.: Studies on cell aggregation: Demonstration of materials with selective cell-binding activity. Proc. Natl. Acad. Sci. USA 49, 742-747 (1963)
Müller, W.E.G.: Cell membranes in sponges. Int. Rev. Cytol. 77, 129-181 (1982)
Müller, W.E.G., Zahn, R.K.: Purification and characterization of a species-specific aggregation factor in sponges. Exp. Cell Res. 80, 95-104 (1973)
Müller, W.E.G., Müller, I., Zahn, R.K.: Two different aggregation principles in reaggregation process of dissociated sponge cells (Geodia cydonium). Experientia 30, 399-902 (1974)
Müller, W.E.G., Müller, I., Zahn, R.K., Kurelec, B.: Species-specific aggregation factor in sponges. VI. Aggregation receptor from the cell surface. J. Cell Sci. 21, 227-241 (1976)
Müller, W.E.G., Müller, I., Pondeljak, V., Kurelec, B., Zahn, R.K.: Species-specific aggregation factor in sponges. Isolation, purification and characterization of the aggregation factor from Suberites domuncula. Differentiation 10, 45-53 (1978a)
Müller, W.E.G., Zahn, R.K., Kurelec, B., Müller, I.: Species-specific aggregation factor in sponges. Nature and alteration of the cell surface charge. Wilhelm Roux' Arch. Entwicklungsmech. Org. 184, 29-40 (1978b)
Müller, W.E.G., Zahn, R.K., Kurelec, B., Müller, I.: Species-specific aggregation factor in sponges. Transfer of a species-specific aggregation factor from Suberites domuncula to cells from Geodia cydonium. Differentiation 10, 55-60 (1978c)
Müller, W.E.G., Zahn, R.K., Kurelec, B., Uhlenbruck, G., Vaith, P., Müller, I.: Aggregation of sponge cells. Glycosyltransferases associated with the aggregation factor. Hoppe-Seyler's Z. Physiol. Chem. 359, 529-537 (1978d)
Müller, W.E.G., Kurelec, B., Zahn, R.K., Müller, I., Vaith, P., Uhlenbruck, G.: Aggregation of sponge cells. Function of a lectin in its homologous biological system. J. Biol. Chem. 254, 7479-7481 (1979a)
Müller, W.E.G., Zahn, R.K., Kurelec, B., Müller, I., Vaith, P., Uhlenbruck, G.: Aggregation of sponge cells. Isolation and characterization of an inhibitor of aggregation receptor from the cell surface. Eur. J. Biochem. 97, 585-591 (1979b)
Müller, W.E.G., Zahn, R.K., Kurelec, B., Müller, I., Uhlenbruck, G., Vaith, P.: Aggregation in sponge cells. A novel mechanism of controlled intercellular adhesion, basing on the interrelation between glycosyltransferase and glycosidase. J. Biol. Chem. 254, 1280-1287 (1979c)
Müller, W.E.G., Zahn, R.K., Kurelec, B., Müller, I., Vaith, P., Uhlenbruck, G.: Aggregation of sponge cells. Species-specific reactivity of a lectin from the sponge Geodia cydonium. Int. J. Biol. Macromol. 2, 297-301 (1980)
Müller, W.E.G., Bernd, A., Zahn, R.K., Kurelec, B., Dawes, K., Müller, I., Uhlenbruck, G.: Xenograft rejection in marine sponges. Isolation and purification of an inhibitory aggregation factor from Geodia cydonium. Eur. J. Biochem. 116, 573-579 (1981a)
Müller, W.E.G., Zahn, R.K., Müller, I., Kurelec, B., Uhlenbruck, G., Vaith, P.: Cell aggregation of the marine sponge Geodia cydonium. Identification of lectin-producing cells. Eur. J. Cell Biol. 24, 28-35 (1981b)
Müller, W.E.G., Conrad, J., Pondeljak, V., Steffen, R., Zahn, R.K.: Electron microscopical characterization of sponge aggregation factors. Tissue Cell 14, 219-223 (1982a)
Müller, W.E.G., Zahn, R.K., Bernd, A., Dawes, K., Müller, I., Uhlenbruck, G., Dorn, A.: Molecular basis of cell-cell recognition: Isolation, synthesis, and function of an aggregation complex from the sponge Geodia cydonium. In: Cellular Recognition (eds. W.A. Frazier, L. Glaser, D.J. Gottlieb), pp. 806-816. New York: Liss 1982b

Müller, W.E.G., Zahn, R.K., Conrad, J., Kurelec, B., Uhlenbruck, G.: Aggregation of sponge cells: Stage dependent, distinct adhesion mechanisms in Cliona celata. Eur. J. Cell Biol. 28, 243-250 (1982c)
Müller, W.E.G., Maidhof, A., Zahn, R.K., Müller, I.: Histoincompatibility reactions in the hydrocoral Millepora dichotoma. Coral Reefs 1, 237-241 (1983a)
Müller, W.E.G., Conrad, J., Schröder, C., Zahn, R.K., Kurelec, B., Dreesbach, K., Uhlenbruck, G.: Characterization of the trimeric, self-recognizing Geodia cydonium lectin I. Eur. J. Biochem. 133, 263-267 (1983b)
Müller, W.E.G., Müller, I., Zahn, R.K., Kurelec, B., Batel, R., Uhlenbruck, G.: Systematic value of histocompatibility reactions in the establishment of closely related sponge species. Description of Geodia rovinjensis. Senckenbergiana Biol. 63, 303-311 (1983c)
Müller, W.E.G., Conrad, J., Schröder, H.C., Zahn, R.K., Kljajic, Z., Müller, I., Uhlenbruck, G.: Cell-cell recognition system in Gorgonians: Description of the basic mechanism. Mar. Biol. 76, 1-6 (1983d)
Müller, W.E.G., Conrad, J., Zahn, R.K., Gramzow, M., Kurelec, B., Uhlenbruck, G.: Identification and isolation of the primary aggregation factor from the cell membrane of the sponge Geodia cydonium. Molec. Cell. Biochem., in press (1985)
Roseman, S.: The synthesis of complex carbohydrates by multiglycosyltransferase systems and their potential function in intercellular adhesion. Chem. Phys. Lipids 54, 217-221 (1970)
Rothbard, J.B., Brackenbury, R., Cunningham, B.A., Edelman, G.M.: Differences in the carbohydrate structures of neural cell-adhesion molecules from adult and embryonic chicken brains. J. Biol. Chem. 257, 11064-11069 (1982)
Sperry, R.W.: Chemoaffinity in the orderly growth of nerve fiber patterns and connections. Proc. Natl. Acad. Sci. USA 50, 703-710 (1963)
Springer, W.R., Haywood, P.L., Barondes, S.H.: Endogenous cell surface lectin in Dictyostelium: Quantitation, elution by sugar, and elicitation by divalent immunoglobulin. J. Cell Biol. 87, 682-690 (1980)
Theodor, J.L.: Distinction between "self" and "not-self" in lower invertebrates. Nature (London) 227, 690-692 (1970)
Tonegawa, Y.: Cell aggregation factor and endogenous lectin in sea urchin embryos. Cell Differ. 11, 335-337 (1982)
Turner, R.S.: Sponge cell adhesions. In: Receptors and Recognition (ed. D.R. Garrod), Ser. B, pp. 199-232. London: Chapman and Hall 1978
Turner, R.S., Burger, M.M.: Involvement of a carbohydrate group in the active site for surface guided reassociation of animal cells. Nature (London) 244, 509-510 (1973)
Uhlenbruck, G.: Immunbiologie, eine Einführung. München: Goldmann 1971
Urushihara, H., Ozaki, H.S., Takeichi, M.: Immunological detection of cell surface components related with aggregation of chinese hamster and chick embryonic cells. Dev. Biol. 70, 206-216 (1979)
Vaith, P., Müller, W.E.G., Uhlenbruck, G.: On the role of D-glucuronic acid in the aggregation of cells from the marine sponge Geodia cydonium. Dev. Comp. Immunol. 3, 259-275 (1979a)
Vaith, P., Uhlenbruck, G., Müller, W.E.G., Holz, G.: Sponge aggregation factor and sponge hemagglutinin. Possible relationships between two different molecules. Dev. Comp. Immunol. 3, 399-416 (1979b)
Weinbaum, G., Burger, M.M.: A two-component system for surface guided reassociation of animal cells. Nature (London) 244, 510-512 (1973)
Wilson, H.V.: On some phenomena of coalescence and regeneration in sponges. J. Exp. Zool. 5, 245-258 (1907)

Subject Index

www.ingramcontent.com/pod-product-compliance
Lightning Source LLC
Chambersburg PA
CBHW061211240926
51846CB00006B/1192
* 9 7 8 3 6 4 2 7 0 2 0 5 1 *